EXPERIMENTS
IN
PHYSICS

A LABORATORY MANUAL
FOR SCIENTISTS AND ENGINEERS

Daryl W. Preston

California State University, Hayward

JOHN WILEY & SONS
New York   Chichester   Brisbane   Toronto   Singapore

TO THE INSTRUCTOR

Experiments in Physics is for a calculus-based, introductory physics course. It consists of an Introduction followed by thirty-two experiments. The experiments follow the order of topics in traditional texts: Mechanics, Heat, Electricity, Magnetism, Optics, and Modern Physics. Each experiment is divided into four major sections:

1. Apparatus - A list of equipment.
2. Introduction - A partial development of the appropriate theoretical concepts.
3. Outcomes - A list of the primary objectives.
4. Experiment - Directions for the experimental procedure with questions included.

Many of the experiments have optional parts which consist of measurements, qualitative observations, and/or calculations. The optional sections are included to provide flexibility for individual instructors. Some instructors may prefer to include optional material and perhaps omit other material; e.g., the optional material in Experiment 7, Rotational Dynamics, may be used as a complete experiment in addition to or in place of the primary experiment.

The Introduction contains six sections:

1. Objectives of the physics laboratory.
2. Laboratory notebook.
3. Error analysis.
4. Significant figures.
5. Graphical analysis.
6. Responsibility of the experimentalist.

Students will not find the concepts presented in the Introduction easy to grasp. It is recommended that the instructor spend some time, especially during the first four or five experiments, discussing this material. It is pointed out in the statement To The Student that they should strive to integrate all of the ideas presented in the Introduction into their laboratory work by the end of the fourth or fifth experiment.

Each experiment is intended for a three-hour laboratory period. However, students will find it difficult to complete some of the experiments in this time interval. If students strive to complete all parts of an experiment, they often sacrifice understanding. Perhaps each student should be urged to work at his or her own pace and strive for understanding first and completion second.

The equipment required for these experiments is reasonably modern and standard, and the following has been done to minimize equipment incompatibilities and to assist instructors in identifying such incompatibilities:

(1) Each experiment includes a list of apparatus.
(2) Experiments using air tracks are described using both spark-timer and digital photo-timer.
(3) Photographs of some equipment, such as rotational dynamics apparatus, oscilloscope, oscillator, multimeter, Geiger-Mueller tube, and scaler/timer are included and specified as typical apparatus.
(4) Suggested values are given for many pieces of apparatus, e.g., 10 Ω resistor, +5 cm focal length lens, etc.

Items (3) and (4) notify students that the apparatus to be used in the laboratory may differ from that specified in the manual. The detail specified in (4) is given to save time for instructors.

TO THE STUDENT

Careful observation is a common and important way of learning. An astronomer might observe the period of radiation pulses emitted by a rotating neutron star in order to gain an understanding of the mechanism producing the pulses. Often times it is desirable to do more than just observe things as they happen, e.g., solid state physicists, working in laboratories, deliberately manipulated lunar samples in order to understand their physical properties. Controlled observation is often called experimentation.

Experimentation primarily consists of careful observations or measurements, recording the data in an orderly form, followed by data analysis, and finally drawing conclusions. Some details as to how one carries out experimentation are discussed in the Introduction. The Introduction is not trivial and you should not expect full comprehension after reading it once. With the help of your instructor you should strive to master the Introduction and to fully integrate the concepts presented there into your laboratory work by the end of the fourth or fifth experiment.

The primary goal of the laboratory is to provide a foundation in experimental science so that you may ultimately carry out independent research.

I urge you to take laboratory work seriously, be prepared before coming to the laboratory, and carry out your work in a careful and thorough manner. The knowledge and skills which you gain from laboratory work will serve you well in many of your future endeavors.

Best wishes,

Daryl W. Preston

ACKNOWLEDGEMENTS

I am indebted to R. H. Good for providing constructive criticism and invaluable suggestions. The assistance he provided during the preparation of this manuscript was very much appreciated.

I would also like to thank Ann Cambra for typing the manuscript.

Any suggestions or comments from students or instructors will be appreciated and may be sent to:

Daryl W. Preston
Physics Department
California State University, Hayward
Hayward, CA 94542

CONTENTS

Introduction

Experiments

EXPERIMENTS IN PHYSICS

A LABORATORY MANUAL
FOR SCIENTISTS AND ENGINEERS

OBJECTIVES OF THE PHYSICS LABORATORY

The basic aims of the laboratory are to have the student:

1. Gain an understanding of some basic physical concepts and theories.

2. Gain familiarity with a variety of instruments and to learn to make reliable measurements.

3. Learn how precisely a measurement can be made with a given instrument and the size of the measurement error. See ERROR ANALYSIS.

4. Learn how to do calculations so that the results have the appropriate number of significant figures. See SIGNIFICANT FIGURES.

5. Learn how to analyze data by calculations and by plotting graphs that illustrate functional relations. See GRAPHICAL ANALYSIS.

6. Learn how to keep an accurate and complete laboratory notebook. See LABORATORY NOTEBOOK.

7. Ultimately, learn how best to approach a new laboratory problem.

LABORATORY NOTEBOOK

In general loose-leaf paper is not appropriate for recording data or doing laboratory calculations. It is recommended that you record data and do calculations directly in your laboratory notebook.

A general question to be considered when writing a lab notebook is: "If I pick up this notebook in a year or two, is there enough information in it for me to understand what was done, why, and what the results and conclusions were? Could I reproduce the experiment if I wanted to?"

With the above in mind, each experiment in your lab notebook should contain the following 8 items:

1. Begin with title, date, partner(s), and number pages at upper right.

2. Purpose: At the beginning state a general purpose in one sentence or two. Throughout the exercise indicate the purpose of each new set of measurements or calculations. (This may simply be a statement of exactly what is being measured if the "why" is obvious.)

3. Sketch the apparatus, free handed, but with parts labeled.

4. Data: All original data are to be recorded directly in the laboratory notebook where all can see them, not on scratch paper. The original data readings are the most important piece of information you have, and their loss should not be risked by recording them on scratch paper. Copying the data wastes valuable time and risks mistakes. Be sure to indicate clearly what is being measured and in what units. You may cross out data that appear to be useless or wrong, but do not erase them - they may turn out to be valuable.

5. Measured quantities should include a figure of uncertainty or "error." See ERROR ANALYSIS.

6. Calculations: All calculated values should be written in the notebook and the method of calculation clearly indicated. They will usually come near the data and may be presented in the form of a table. Each calculated result should include appropriate significant figures. See SIGNIFICANT FIGURES.

7. Graphs: Follow the guidelines listed in GRAPHICAL ANALYSIS.

8. Results and conclusion. Tell briefly what you did and how it came out. If you measured a physical constant, how does it compare with the "accepted" value in the light of your estimated errors?

The format is not rigid, but should follow the order in which you worked. The method of approach is often more important than obtaining a particular result. Please do not create literature; your statement of purpose and your remarks pertaining to results and conclusion should be concise.

ERROR ANALYSIS

Often the verification of a physical law or the determination of a physical quantity involves measurements. A reading taken from the scale on a voltmeter, a stopclock or a meter stick, for example, may be directly related by a chain of analysis to the quantity or law under study. Any uncertainty in these readings would result in an uncertainty in the final result. A measurement by itself, without a quantitative statement as to the uncertainty involved, is of limited usefulness. It is therefore essential that any course in basic laboratory technique include a discussion of the nature of the uncertainty in individual measurements and the manner in which uncertainties in two or more measurements are propagated to determine the uncertainty in the quantity or law being investigated. Such uncertainties are often called experimental errors.

Types of Experimental Errors

In the collection of data two types of experimental errors, systematic errors and random errors usually contributed to the error in the measured quantity.

Systematic errors are due to identifiable causes and can, in principle, be eliminated. Errors of this type result in measured values which are consistently too high or consistently too low. Systematic errors may be of four kinds:

a. Instrumental - e.g., a poorly calibrated instrument such as a thermometer that reads 102 °C when immersed in boiling water and 2 °C when immersed in ice water at atmospheric pressure. Such a thermometer would result in measured values which are consistently too high.

b. Observational - e.g., parallax in reading a meter scale.

c. Environmental - e.g., an electrical power "brown out" which causes measured currents to be consistently too low.

d. Theoretical – due to simplifactions of the model system or approximations in the equations describing it, e.g., if a frictional force is acting during the experiment but such a force is not included in the theory, then the theoretical and experimental results will consistently disagree.

In principle an experimentalist wants to identify and eliminate systematic errors.

<u>Random errors</u> are positive and negative fluctuations which cause about half of the measurements to be too high and half too low. Sources of random errors cannot always be identified. Possible sources of random errors are:

a. Observational – e.g., errors in judgment of an observer when reading the scale of a measuring device to the smallest division.

b. Environmental – e.g., unpredictable fluctuations in line voltage, temperature, or mechanical vibrations of equipment.

Random errors, unlike systematic errors, can often be quantified by statistical analysis; therefore, the effects of random errors on the quantity or physical law under investigation can often be determined.

The distinction between random errors and systematic errors can be illustrated with the following example. Suppose the measurement of a physical quantity is repeated five times under the same conditions. If there are only random errors, then the five measured values will be spread about the "true value"; some will be too high and others too low as shown in Figure 1a. If in addition to the random errors there is also a systematic error, then the five measured values will be spread, not about the true value, but about some displaced value as shown in Figure 1b.

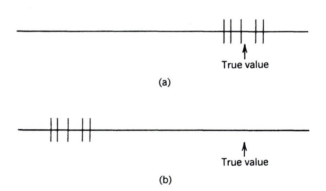

Figure 1. Set of measurements (a) with random errors only and (b) with systematic and random errors. Each mark indicates the result of a measurement.

Statistical Analysis of Random Errors

If a physical quantity, such as a length measured with a meter stick or a time interval measured with a stopcock, is measured many times, then a distribution of readings is obtained due to random errors. For such a set of data the average or mean value $\bar{x}$ is defined by

$$\bar{x} = \frac{1}{n} \sum_{i=1}^{n} x_i \tag{1}$$

where x_i is the i th measured value and n is the total number of measurements.

The n measured values will be distributed about the mean value as shown in Figure 2. (In many cases $\bar{x}$ approaches the "true value" if n is very large and there are no systematic errors.) A small spread of measured values about the mean value implies high <u>precision</u>.

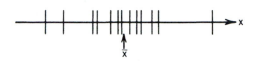

Figure 2. Each mark indicates the result of a measurement. The measured values are distributed about the mean value $\bar{x}$.

Now that we have determined the "best value" for the measurement, i.e., $\bar{x}$, we need to estimate the uncertainty or error in this value. We start with defining one way in which the spread of data about the mean value can be characterized.

The <u>standard deviation</u> s is defined as

$$s = \sqrt{\frac{1}{n-1} \sum_{i=1}^{n} (x_i - \bar{x})^2} \tag{2}$$

If the standard deviation is small, then the spread in the measured values about the mean is small, hence the precision in the measurements is high. Note the standard deviation is always positive and it has the same units as the measured values.

The error or uncertainty in the mean value, $\bar{x}$, is the <u>standard deviation of the mean</u>, s_m, which is defined to be:

$$s_m = \frac{s}{\sqrt{n}} \tag{3}$$

where s is the standard deviation and n is the total number of measurements.

The result to be reported is then

$$\bar{x} \pm s_m \tag{4}$$

The interpretation of Equation (4) is that the measured value probably lies in the range from $\bar{x} - s_m$ to $\bar{x} + s_m$.

In order to more fully discuss the spread or distribution of measured values about the mean value it is useful to consider the normal or Gauss distribution.

Normal or Gauss Distribution

Figure 3a shows the distribution of measured values about the mean value for two sets of data, where each set is n repeated measurements of the same physical quantity. The x axis has been divided into increments of width Δx and each dot indicates a measured value. The dots are spread vertically for clarity. The data on the left is more closely clustered about the mean; hence it implies higher precision. In Figure 3b the number of measured values, N(x), in an increment Δx centered on x is plotted vertically. Note the curve on the left, which corresponds to the data of higher precision, is more sharply peaked than the curve on the right. The smooth curves which are nonsymmetrical, in Figure 3b, are drawn to illustrate the approximate dependence of the number of measured values N(x) on x. If the number of measurements n becomes very large, then the measured values are symmetrically distributed about the mean value as shown in Figure 3c. For very large n the standard deviation is denoted by σ. Each curve in 3c, N(x) vs. x, represents the frequency with which the value x is obtained as the result of any single measurement. Ideally, the analytical expression for such curves is

$$N(x) = \frac{n}{\sqrt{2\pi}\,\sigma}\, e^{-(x-\bar{x})^2/2\sigma^2} \tag{5}$$

where n is the very large number of measurements, $\bar{x}$ is the mean value, and σ is the standard deviation. Equation (5) is the normal or Gauss distribution.

For a very large number of measurements n, the normal or Gauss distribution is the theoretical distribution of measured values x about the mean value $\bar{x}$. If the measurements are carried out with high precision, then σ will be small and the normal or Gauss distribution will be sharply peaked at the mean value $\bar{x}$. See Figure 4.

If both sides of Equation (5) are divided by n and defining N(x)/n to be P(x), we have

$$P(x) = \frac{1}{\sqrt{2\pi}\,\sigma}\, e^{-(x-\bar{x})^2/2\sigma^2} \tag{6}$$

The probability of obtaining the value x as a result of any single measurement is given by P(x). Note that the most probable value resulting from any single measurement is the mean value $\bar{x}$.

The Gauss distribution has the property that 68% of the measurements will fall within the range from

$$\bar{x} - \sigma \quad \text{to} \quad \bar{x} + \sigma \tag{7}$$

and 95% of the measurements will fall within the range from

$$\bar{x} - 2\sigma \quad \text{to} \quad \bar{x} + 2\sigma \tag{8}$$

Young, Hugh D., Statistical Treatment of Experimental Data, McGraw-Hill, 1962.

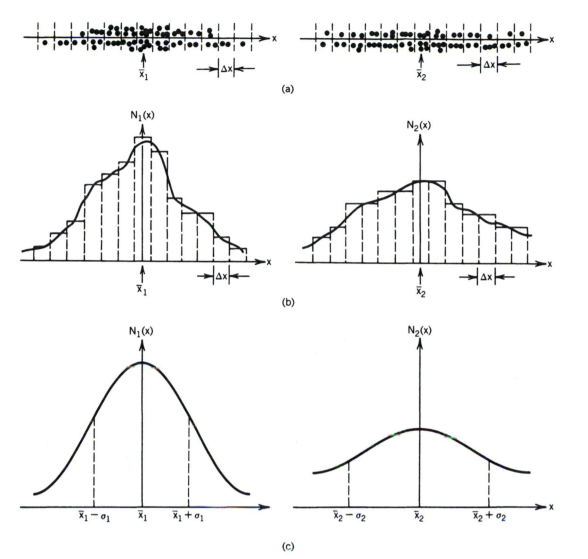

Figure 3.   Two sets of measurements for the same physical quantity.
(a) Each dot indicates the result of a measurement. The
dots are spread vertically for clarity.   (b) N(x) is the
number of measured values in an increment Δx centered
on x.   (c) For very large n the distribution of measured
values about the mean value is the normal or Gauss
distribution, and $\bar{x}_1$ and $\bar{x}_2$ approach the same value, the
"true value."   The data of higher precision has a smaller
standard deviation, i.e., $\sigma_1 < \sigma_2$.

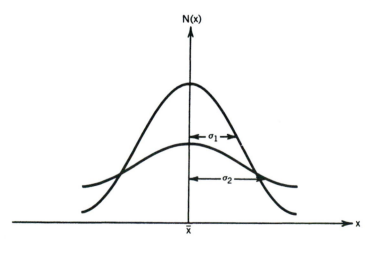

Figure 4.   Gauss distributions for
the same $\bar{x}$ and two values
of σ. σ_1 is smaller than
σ_2 and implies data of
higher precision.

Summarizing the best way to treat random errors:

a. Repeat the measurement n times. The measurement could be, for example, a reading of an ammeter to measure electrical current or a reading of a stopclock to measure a time interval.

b. Calculate the mean value $\bar{x}$, the standard deviation s, and the standard deviation of the mean s_m.

c. In the limit when $n \to \infty$, then $\bar{x} \to$ "true value," $s \to \sigma$ and $s_m \to \sigma_m$.

A word of caution is in order pertaining to the distribution of measured values about the mean value. The distribution of measurements is often describable by the Gauss distribution; however, not all distributions follow the Gauss distribution. In the preceding discussion we have assumed the measurements are describable by the Gauss distribution, as often is the case.

Review Exercises

For each exercise calculate: (a) the mean value, (b) the standard deviation s (c) the standard deviation of the mean s_m, (d) the percentage error, and (e) the result to be reported.

1. The data are a set of measurements of the length of a sheet of paper, made with a 30-cm rule.

$\ell_1 = 27.94$ cm $\ell_2 = 27.96$ cm $\ell_3 = 27.99$ cm $\ell_4 = 27.97$ cm

$\ell_5 = 28.00$ cm $\ell_6 = 27.93$ cm $\ell_7 = 27.96$ cm $\ell_8 = 27.98$ cm

Answers: (a) $\bar{\ell} = 27.97$ cm   (d) $\dfrac{s_m}{\bar{\ell}} \times 100\% = 0.04\%$

(b) $s = 0.02$ cm

(c) $s_m = 0.01$ cm   (e) $\bar{\ell} \pm s_m = 27.97 \pm 0.01$ cm

We round off the calculated values, keeping an appropriate number of significant figures. See SIGNIFICANT FIGURES.

2. The data are a set of measurements of a person's weight, made with bathroom scales.

$W_1 = 135.0$ lbs $W_2 = 136.5$ lbs $W_3 = 134.0$ lbs $W_4 = 134.5$ lbs

Answers: (a) $\bar{W} = 135.0$ lbs   (d) $\dfrac{s_m}{\bar{W}} \times 100\% = 0.4\%$

(b) $s = 1.1$ lbs

(c) $s_m = 0.6$ lbs   (e) $\bar{W} \pm s_m = 135.0 \pm 0.6$ lbs

3. The data are a set of measurements of the period of a pendulum, made with the second hand of a watch.

$t_1 = 2.1$ s $t_2 = 2.5$ s $t_3 = 2.3$ s

Answers: (a) $\bar{t}$ = 2.3 s   (d) $\dfrac{s_m}{\bar{t}}$ × 100% = 4%

(b) s = 0.2 s

(c) s_m = 0.1 s   (e) $\bar{t} \pm s_m$ = 2.3 ± 0.1 s

In general we won't use the above-described best way to determine random error because we seldom have the time and resources to make so many measurements, and because a simpler method of estimation of random error to be described below is usually adequate.

Estimation of Random Error

We will estimate measurement errors in a somewhat subjective way based on judgment and experience. For example, we know that the ERROR IN A GIVEN INSTRUMENT IS LIKELY TO BE ABOUT THE SAME SIZE AS THE SMALLEST SEPARATE DIVISION ON THAT INSTRUMENT; but instruments vary widely in the reliability of that smallest division, and some judgment must come into play. If we measure the position of a mark to be 92.4 cm, using a meter stick whose smallest division is a millimeter, then we might write the result as 92.4 ± 0.1 cm.

Figure 5 illustrates one way judgment must come into play when estimating measurement errors. The distance d_1 is the separation of the sharply defined vertical lines, and d_2 is the distance from center to center of the two "globs." Even though we may measure d_1 and d_2 with the same ruler (hence smallest separate division is the same) the error in d_2 is likely to be larger than that of d_1. Measure d_1 and d_2 with a ruler having a smallest division of 0.1 cm and estimate the errors!

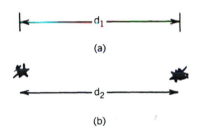

Figure 5. The error in d_2 is likely to be larger than that in d_1 since the centers of the globs are not sharply defined.

Propagation of Errors

Propagation of errors is simply a method to determine the error in a value where the value is calculated using two or more measured values with known estimated errors. This method will be discussed separately for addition and subtraction of measurements, and for multiplication and division of measurements.

ADDITION AND SUBTRACTION OF MEASUREMENTS

Suppose x, y, and z are three measured values and the estimated errors are δx, δy, and δz.

The results of the three measurements would be reported in the form:

12

$$x \pm \delta x$$
$$y \pm \delta y \qquad (9)$$
$$z \pm \delta z$$

where each estimated error may be the SMALLEST DIVISION OF THE MEASURING INSTRU-MENT.

If w, a value to be calculated from the measurements, is defined to be:

$$w = x + y - z \qquad (10)$$

then knowing the estimated errors in x, y and z, what is the error in w, δw? We could try to answer this question by calculating the differential of both sides of Equation (10):

$$dw = dx + dy - dz \qquad (11)$$

We assume the measurements are carried out such that the estimated errors are much smaller than the measurements:

$$\delta x << x$$
$$\delta y << y \qquad (12)$$
$$\delta z << z$$

Thus we may approximate the differentials in Equation (11) with "deltas" i.e.,

$$\delta w = \delta x + \delta y - \delta z \qquad \text{(incorrect)} \qquad (13)$$

Now the estimated errors are random and hence are equally likely to be too high or too low, i.e., positive or negative. Therefore, we cannot assume the error in z will always subtract from the errors in x and y. Hence we change the negative sign in Equation (13):

$$\delta w = \delta x + \delta y + \delta z \qquad \text{(unacceptable)} \qquad (14)$$

Statistical analysis shows that a better approximation is to write δw as the square root of the sum of estimated errors squared:

$$\delta w = \sqrt{(\delta x)^2 + (\delta y)^2 + (\delta z)^2} \qquad (15)$$

If one of the estimated errors is significantly larger than the others, then we may ignore the others; for example, if δy is the largest, then:

$$\delta w \simeq \sqrt{(\delta y)^2} = \delta y \qquad (16)$$

Thus for most laboratory work we recommend:

WHEN MEASUREMENTS ARE ADDED OR SUBTRACTED, THE LARGEST ERROR IN THE MEASUREMENT BECOMES THE ERROR OF THE RESULT.

EXAMPLES

1. Suppose three measured lengths and their estimated errors are:

$$\ell_1 \pm \delta\ell_1 = 23.5 \pm 0.1 \text{ cm}$$
$$\ell_2 \pm \delta\ell_2 = 17.8 \pm 0.4 \text{ cm}$$
$$\ell_3 \pm \delta\ell_3 = 93.9 \pm 0.2 \text{ cm}$$

If the quantity to be calculated, L, is defined to be:

$$L = \ell_1 + \ell_2 - \ell_3$$

then

$$L \pm \delta L = -52.6 \pm 0.4 \text{ cm}$$

since 0.4 cm is the largest error.

2. Suppose two measured masses and their estimated errors are:

$$m_1 \pm \delta m_1 = 1.746 \pm 0.010 \text{ kg}$$
$$m_2 \pm \delta m_2 = 0.507 \pm 0.010 \text{ kg}$$

The quantity to be calculated, M, is defined to be:

$$M = m_1 + m_2$$

Since the errors, δm_1 and δm_2 are the same, then

$$\delta M = \sqrt{(0.010)^2 + (0.010)^2} = \sqrt{2} \times 0.010 = 0.014 \text{ kg}$$

and

$$M \pm \delta M = 2.253 \pm 0.014 \text{ kg}$$

3. Suppose two time intervals and their errors are:

$$t_1 \pm \delta t_1 = 0.743 \pm 0.005 \text{ s}$$
$$t_2 \pm \delta t_2 = 0.384 \pm 0.005 \text{ s}$$

The total time t is defined to be:

$$t = 2t_1 + 5t_2$$

Perhaps it is worthwhile to "derive" δt, since t_1 and t_2 are multiplied by constants. Calculating the differential of t:

$$dt = 2dt_1 + 5dt_2$$

Approximately the differentials with "deltas"

$$\delta t = 2\delta t_1 + 5\delta t_2$$

Statistical analysis shows a better approximation is

$$\delta t = \sqrt{(2\delta t_1)^2 + (5\delta t_2)^2}$$

Substituting numerical values:

$$\delta t = \sqrt{(0.010)^2 + (0.025)^2} \approx 0.025 \text{ s}$$

where we approximated the right-hand side by the larger error. The final result is

$$t \pm \delta t = 1.127 \pm 0.025 \text{ s}$$

IMPORTANT POINT: IN CALCULATING δt, WE MUST MULTIPLY δt_1 AND δt_2 BY THE SAME FACTORS WHICH MULTIPLY t_1 AND t_2 IN THE EQUATION DEFINING t.

MULTIPLICATION AND DIVISION OF MEASUREMENTS

The area A of a rectangle of width w and height h is w·h. Suppose we measure the width and height, and estimate their errors. Then we know $w \pm \delta w$ and $h \pm \delta h$, and we want to calculate $A \pm \delta A$.

To determine δA, we first use differential calculus to obtain the differential area dA:

$$dA = \frac{dA}{dw} dw + \frac{dA}{dh} dh$$

$$= h \, dw + w \, dh \tag{17}$$

Dividing both sides by A, where on the right-hand side we write A as w·h, we obtain

$$\frac{dA}{A} = \frac{dw}{w} + \frac{dh}{h} \tag{18}$$

If the estimated errors are much smaller than their measured values, we may replace the differentials with "deltas":

$$\frac{\delta A}{A} = \frac{\delta w}{w} + \frac{\delta h}{h} \tag{19}$$

Note that each term in Equation (19) is a ratio of error to measured value, i.e., each term is a dimensionless fractional error.

As before, statistical analysis shows that a better estimation of the fractional error in area, $\delta A/A$, is:

$$\frac{\delta A}{A} = \sqrt{\left(\frac{\delta w}{w}\right)^2 + \left(\frac{\delta h}{h}\right)^2} \tag{20}$$

We may estimate the fractional error in the area by ignoring the smaller fractional error in the measured values, e.g., if $\delta h/h$ is larger than $\delta w/w$, then

$$\frac{\delta A}{A} \approx \sqrt{\left(\frac{\delta h}{h}\right)^2} = \frac{\delta h}{h} \qquad (21)$$

Thus δA is

$$\delta A \approx A\,\frac{\delta h}{h} \qquad (22)$$

> WHEN MEASUREMENTS ARE MULTIPLIED OR DIVIDED, THE LARGEST FRACTIONAL ERROR IN THE MEASUREMENTS BECOMES THE FRACTIONAL ERROR IN THE RESULT.

EXAMPLES

1. We calculate a gravitational potential energy U and its error δU where U = mgh.

   If $\quad m \pm \delta m = 0.1000 \pm 0.0005$ kg $\qquad \frac{\delta m}{m} = 5 \times 10^{-3}$

   and $\quad g \pm \delta g = 9.80 \pm 0.04$ ms^{-2} $\qquad \frac{\delta g}{g} = 4 \times 10^{-3}$

   and $\quad h \pm \delta h = 0.689 \pm 0.002$ m $\qquad \frac{\delta h}{h} = 3 \times 10^{-3}$

   then $\quad U = mgh = 0.6752$ joules $\qquad \frac{\delta u}{u} \approx \frac{\delta m}{m} = 5 \times 10^{-3}$

   since the largest fractional error is 5×10^{-3}. The final result should quote error, not fractional error. The error is:

   $$\delta U = 5 \times 10^{-3} \times 0.6752 = 0.003376 \text{ joules}$$

   We round off the final result and its error to a consistent number of significant figures, and we give the error to one (or at most two) figures. Thus the final result becomes:

   $$U \pm \delta U = 0.675 \pm 0.003 \text{ joules}$$

2. We calculate the speed v from the measured distance x and time t, where v = x/t. Perhaps it is worthwhile to "derive" the error in v, δv. Calculating the differential of v, dv:

   $$dv = \frac{dv}{dx}\,dx + \frac{dv}{dt}\,dt$$

   $$= \frac{1}{t}\,dx - \frac{x}{t^2}\,dt \qquad (23)$$

   Approximating differentials with "deltas"

16

$$\delta v = \frac{\delta x}{t} - \frac{x}{t^2}\, \delta t \qquad (24)$$

Dividing both sides by v, where on the right-hand side we write v as x/t

$$\frac{\delta v}{v} = \frac{\delta x}{x} - \frac{\delta t}{t} \qquad (25)$$

We cannot assume the fractional error in time will always subtract from the fractional error in distance, hence we change the sign of the fractional error in time:

$$\frac{\delta v}{v} = \frac{\delta x}{x} + \frac{\delta t}{t} \qquad (26)$$

Statistical analysis shows a better approximation of $\frac{\delta v}{v}$ is

$$\frac{\delta v}{v} = \sqrt{\left(\frac{\delta x}{x}\right)^2 + \left(\frac{\delta t}{t}\right)^2} \qquad (27)$$

If $\frac{\delta t}{t} > \frac{\delta x}{x}$, then we ignore the smaller fractional error and

$$\frac{\delta v}{v} \approx \frac{\delta t}{t} \qquad (28)$$

Compare Equations (23) – (28) with Equations (17) – (22).

If $\qquad x \pm \delta \dot x = 0.63 \pm 0.02$ cm $\qquad \frac{\delta x}{x} = 0.03$

and $\qquad t \pm \delta t = 1.71 \pm 0.10$ s $\qquad \frac{\delta t}{t} = 0.06$

then $\qquad v = \frac{x}{t} = 0.368$ ms^{-1} $\qquad \frac{\delta v}{v} \approx \frac{\delta t}{t} = 0.06$

The error is

$$\delta v \approx 0.06 \times 0.368 = 0.022 \text{ ms}^{-1}$$

Thus the final result becomes

$$v \pm \delta v = 0.37 \pm 0.02 \text{ ms}^{-1}$$

3. The kinetic energy K is defined in terms of the mass M and speed v: $K = 1/2\, Mv^2$.

If $\qquad M \pm \delta M = 0.352 \pm 0.001$ kg $\qquad \frac{\delta M}{M} = 0.003$

and $\qquad v \pm \delta v = 1.91 \pm 0.01$ ms^{-1} $\qquad \frac{\delta v}{v} = 0.005$

then $\qquad K = 1/2\, Mv^2 = 0.642$ joules $\qquad \frac{\delta K}{K} \approx \frac{2\delta v}{v} = 0.010$

The fractional error in velocity is multiplied by 2 since the velocity is squared in the kinetic energy. The error in K is then

$$\delta K \simeq 0.010 \times 0.642 = 0.006 \text{ joules}$$

The final result is

$$K \pm \delta K = 0.642 \pm 0.006 \text{ joules}$$

For more complicated functions it may be easier to calculate extreme values. For example, if $\theta \pm \delta\theta = 10° \pm 2°$ then $\cos 8° = 0.990$, $\cos 10° = 0.985$, and $\cos 12° = 0.978$. Hence $\cos \theta \pm \delta(\cos \theta) = 0.985 \pm 0.006$ where the average of 0.005 and 0.007 was taken.

Percent discrepancy is one method to compare an experimental value with a known accepted or true value. The definition of percent discrepancy is

$$\text{percent discrepancy} = \left| \frac{\text{accepted value} - \text{experimental value}}{\text{accepted value}} \right| \times 100\% \qquad (29)$$

For example, if we measure g as $9.20 \pm 0.20 \text{ ms}^{-2}$ and knowing the accepted value is 9.80 ms^{-2}, then the percent discrepancy is 6.1%. The fact that the accepted value so exceeds the experimental value suggests that something is wrong with the measuring apparatus; perhaps a systematic error was not eliminated. If by contrast our measurement comes out $9.80 \pm 0.20 \text{ ms}^{-2}$, then the percent discrepancy is zero by accident, and this does not imply that our experimental error is zero.

SIGNIFICANT FIGURES

The significant figures in a number are all figures that are obtained directly from the measuring process and exclude those zeros which are included solely for the purpose of locating the decimal point. This definition will be illustrated with a number of examples.

Number	Number of Significant Figures	Remarks
2	1	Implies $\cong$ 25% precision
2.0	2	Implies $\cong$ 2.5% precision
2.00	3	Implies $\cong$ 0.25% precision
0.136	3	Leading zero is not necessary, but it does make the reader notice the decimal point.
2.483	4	
2.483×10^3	4	
310	2 or 3	Ambiguous. The zero may be significant or it may be present only to show the location of the decimal point.
3.10×10^2	3	No ambiguity
3.1×10^2	2	

A measurement and its experimental error should have their last significant digits in the same location (relative to the decimal point). Examples: 54.1 ± 0.1, 121 ± 4, 8.764 ± 0.002, $(7.63 \pm 0.10) \times 10^3$.

Handling of Significant Figures in Calculations

Properly, the correct number of significant figures to which a result should be quoted is obtained via error analysis. However, error analysis takes time, and frequently in actual laboratory practice it is postponed. In such a case, one should retain enough significant figures that round-off error is no danger, but not so many as to constitute a burden. Here is an example:

$$0.91 \times 1.23 = 1.1 \qquad \text{WRONG}$$

In this case, the numbers 0.91 and 1.23 are known to about 1%, whereas the result, 1.1, is defined to about 10%. In this extreme case, the accuracy of the result is reduced by almost a factor of ten, due to round-off error. Now, a factor of ten in accuracy is usually precious and expensive, and it must not be thrown away by careless data analysis.

$$0.91 \times 1.23 = 1.1193 \qquad \text{WRONG}$$

The extra digits, which are not really significant, are just a burden, and in addition they carry the incorrect implication of a result of absurd accuracy.

$$0.91 \times 1.23 = 1.12 \qquad \text{okay}$$
$$0.91 \times 1.23 = 1.119 \qquad \text{less good, but still acceptable}$$

In multiplication or division it is often acceptable to keep the same number of significant figures in the product or quotient as are in the least precise factor. Examples:

$$2.6 \times 31.7 = 82.42 = 82$$
$$5.3 \div 748 = .007085 = 0.0071$$

Handling of significant figures in addition and subtraction will be illustrated with examples:

51.4	7146	20.8	1.4693
− 1.67	− 12.8	18.72	10.18
49.73 → 49.7	7133.2 → 7133	+ .851	+ 1.062
		40.371 → 40.4	12.7113 → 12.71

Note that the answer is appropriately rounded off in each example.

Review Exercises

In each exercise, measured values and their errors are given. A quantity to be calculated is defined in terms of the measured quantities. Calculate: (a) the defined quantity, (b) the error in the defined quantity, and (c) the percentage error in the defined quantity.

1. The quantity to be calculated is L where $L = \ell_1 + \ell_2 - \ell_3$ and the measured values and their errors are:

   $\ell_1 \pm \delta \ell_1 = 17.4 \pm 0.2$ cm

   $\ell_2 \pm \delta \ell_2 = 9.76 \pm 0.05$ cm

   $\ell_3 \pm \delta \ell_3 = 11 \pm 1$ cm

   Answers: (a) $L = 16$ cm

   (b) $\delta L = 1$ cm

   (c) $\dfrac{\delta L}{L} \times 100\% = 6\%$

2. The quantity to be calculated is $\bar{v}$, the average speed, where $\bar{v} = x/t$ and the measured values and their errors are:

   $x \pm \delta x = (1.748 \pm 0.010) \times 10^{-2}$ m

   $t \pm \delta t = (5.41 \pm 0.05) \times 10^{-3}$ s

   Answers: (a) $\bar{v} = 3.23$ ms^{-1}

   (b) $\delta \bar{v} = 0.03$ ms^{-1}

   (c) $\dfrac{\delta \bar{v}}{\bar{v}} \times 100\% = 0.9\%$

3. The quantity to be calculated is K, the kinetic energy, where $K = 1/2\, mv^2$ and the measured values and their errors are:

   $m \pm \delta m = 1.25 \pm 0.05$ kg

   $v \pm \delta v = 0.87 \pm 0.01$ ms^{-1}

   Answers: (a) $K = 0.47$ kg m^2 s^{-2}

   (b) $\delta K = 0.02$ kg m^2 s^{-2}

   (c) $\dfrac{\delta K}{K} \times 100\% = 4\%$

4. The quantity to be calculated is F, the gravitational force, where $F = G\, m_1 m_2 / r^2$ and the measured values and their errors are:

   $m_1 \pm \delta m_1 = 19.7 \pm 0.2$ kg

   $m_2 \pm \delta m_2 = 9.4 \pm 0.2$ kg

   $r \pm \delta r = 0.641 \pm 0.009$ m

   $G = 6.67 \times 10^{-11}$ N m^2 kg^{-2}

   Answers: (a) $F = 3.0 \times 10^{-8}$ N

   (b) $\delta F = 0.1 \times 10^{-8}$ N

   (c) $\dfrac{\delta F}{F} \times 100\% = 3\%$

GRAPHICAL ANALYSIS

It is often desirable to find the relationship between measured variables. One good way to do this is to plot a graph of the data and then analyze the graph. The following guidelines should be followed in plotting your data:

1. Use a sharp pencil or pen. A broad-tipped pencil or pen will introduce unnecessary inaccuracies.

2. Draw your graph on a full page of your notebook. A compressed graph will reduce the accuracy of your graphical analysis.

3. Give the graph a concise title.

4. The dependent variable should be plotted along the vertical (y) axis and the independent variable along the horizontal (x) axis.

5. Label axes and include units.

6. Select a scale for each axis and start each axis at zero, if possible.

7. Use error bars to indicate errors in measurements, e.g.,

$$\text{data point} \longrightarrow \quad \longleftarrow \text{error range}$$

8. Draw a smooth curve through the data points. If the errors are random, then about 1/3 of the points will not lie within their error range of the best curve.

As an example consider the study of the speed of an object (dependent variable) as a function of time (independent variable). The data are:

Speed (m/s)	Time (s)
0.45 ± 0.06	1
0.81 ± 0.06	2
0.91 ± 0.06	3
1.01 ± 0.06	4
1.36 ± 0.06	5
1.56 ± 0.06	6
1.65 ± 0.06	7
1.85 ± 0.06	8
2.17 ± 0.06	9

Following the above guidelines, the data are graphed in Figure 6.

Speed vs. Time

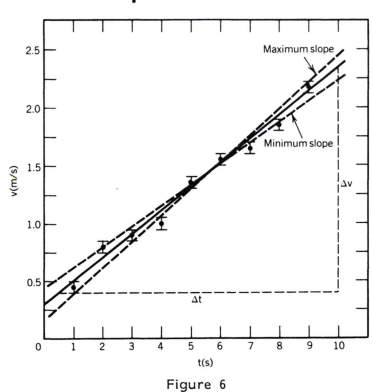

Figure 6

The graphed data shows the speed v is a linear function of the time t. The general equation for a straight line is

$$y = mx + b \tag{30}$$

where m is the slope of the line and b, the vertical intercept, is the value of y when x = 0. Let $v = y$, and $x = t$, $a = m$, and $v_o = b$, then

$$v = at + v_o \tag{31}$$

This is the form of the equation for the solid line drawn through the data, where v_0 is the value of the velocity at $t = 0$ and a is the slope of the line which is the acceleration of the object. From the graph: $v_0 = 0.32$ m/s. To determine the slope select two points on the line which are well separated; then:

$$a = \text{slope} = \frac{\Delta v}{\Delta t} = \frac{2.35 - 0.40 \ (m/s)}{10.0 - 0.5 \ (s)} = \frac{1.95 \ (m/s)}{9.5 \ (s)} = 0.20 \ m/s^2 \qquad (32)$$

The equation for the line is:

$$v = 0.20t + 0.32 \ (m/s) \qquad (33)$$

How accurate are the values for the slope (0.20 m/s^2) and the vertical intercept (0.32 m/s), i.e., what is the uncertainty in the slope and the vertical intercept? In the graph of speed vs. time the solid line represents the best straight line whose slope we already calculated. The two dashed lines represent the lines of greatest and least possible slopes which reasonably fit the data. Here "reasonably" means each line should intercept about 2/3 of the error bars. The error or uncertainty in the slope is defined to be

$$\text{error in slope} = \frac{\text{maximum slope} - \text{minimum slope}}{2}$$

Using δa for the error in a

$$\delta a = (\text{maximum slope} - \text{minimum slope})/2$$
$$= \frac{0.23 - 0.19}{2} \ (m/s^2) = 0.02 \ m/s^2 \qquad (34)$$

The experimental value and error for a are:

$$a \pm \delta a = 0.20 \pm 0.02 \ m/s^2$$

The error in the vertical intercept is found from the vertical intercepts of the lines of maximum and minimum slope:

$$\text{error in vertical intercept} = (\text{vertical intercept of minimum slope line} - \text{vertical intercept of maximum line})/2$$

Using δv_0 for the error in v_0:

$$\delta v_0 = (0.45 - 0.17)/2 \ (m/s) = 0.31 \ m/s$$

The experimental value and error for v_0 are:

$$v_0 \pm \delta v_0 = 0.32 \pm 0.31 \ m/s$$

As a second example consider the study of the distance traveled by an object as a function of time. The data are:

24

Distance (m)	Time (s)
0.20 ± 0.05	1
0.43 ± 0.05	2
0.81 ± 0.05	3
1.57 ± 0.10	4
2.43 ± 0.10	5
3.81 ± 0.10	6
4.80 ± 0.20	7
6.39 ± 0.20	8

The data are graphed, using the above guidelines, in Figure 7.

Distance vs. Time

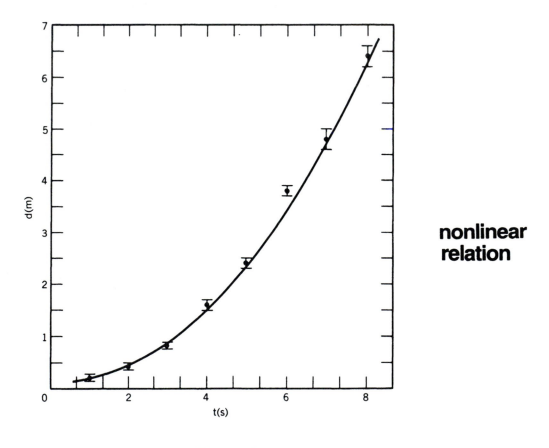

nonlinear relation

Figure 7

In this case a straight line through the data points would not be acceptable, i.e., the graphed data shows the distance d is not a linear function of the time t. Inspection of the graph suggests that d may be proportional to t^n, where n > 1, e.g., d may be a quadratic function of time and hence n = 2.

Suppose we know the theoretical relation between d and t is

$$d = \frac{1}{2} at^2 \qquad (35)$$

where a is the object's acceleration. Often it is useful to know if the data agrees with the theory. If the data follow the above theoretical relation then a graph of d vs. t^2 should result in a straight line.

Distance versus time squared is graphed in Figure 8.

Distance vs. (Time)²

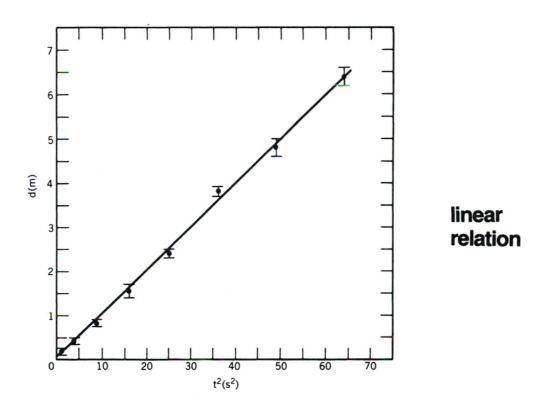

linear relation

Figure 8

The graph indicates d is a linear function of t^2 and hence the data agrees with the theoretical relation. The equation for the straight line is

$$d = mt^2 + d_o \qquad (36)$$

where m is the slope and d_o the vertical intercept.

Plotting Data on Semilog Paper

Often the relationship between the measured variables is not linear. For example, consider the intensity of light, I, transmitted through a sample of thickness x. See Figure 9, where I_o is the intensity of the incident light.

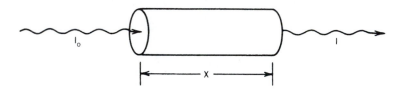

Figure 9

Lambert's law states the theoretical relationship between the dependent variable I and the independent variable x:

$$I = I_o e^{-\mu x} \tag{37}$$

where μ is the absorption coefficient, a constant which depends on the wavelength of light and the absorbing properties of the sample. Suppose I is measured as a function of x, and the data are plotted as shown in Figure 10.

Light Intensity vs. Sample Thickness

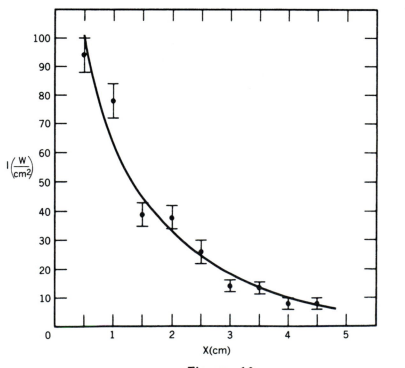

nonlinear relation

Figure 10

From the smooth curve it would be difficult to determine the relationship between I and x, i.e., it would be difficult to conclude that the data obey Lambert's law.

A good way to determine the experimental relationship between I and x is to use semilog paper. Semilog paper has a logarithmic y-axis (it automatically takes logarithms of data plotted) and a regular spaced x-axis. The data are plotted on semilog paper in Figure 11. Note that there is never a zero on the logarithmic axis.

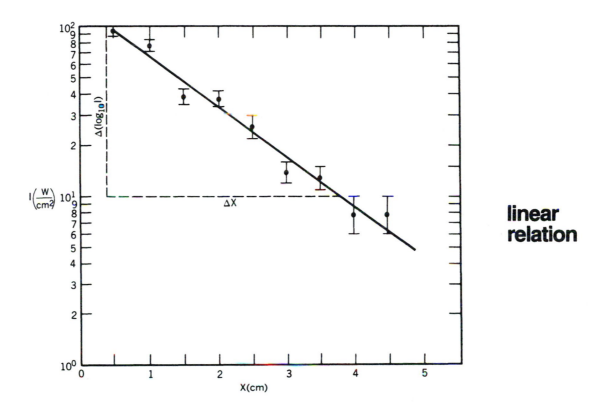

Figure 11

The smooth curve drawn through the data is a straight line with a negative slope and the intensity at the point on the vertical axis intercepted by the curve is I_o. Lambert's law does agree with this result as can be seen by taking the logarithm of Lambert's law:

$$\log_{10} I = \log_{10} (I_o \, e^{-\mu x})$$
$$= \log_{10} e^{-\mu x} + \log_{10} I_o$$
$$= -\mu x \, \log_{10} e + \log_{10} I_o \qquad (38)$$
$$= -0.434 \, \mu x + \log_{10} I_o$$

Again the general equation of a straight line is of the form:

$$y = mx + b \tag{39}$$

where m is the slope and b is the vertical intercept. Now let $y = \log_{10} I$, $m = -.434 \mu$, and $b = \log_{10} I_0$. Then if $\log_{10} I$ is plotted vertically and x is plotted horizontally, the curve will be a straight line with slope $-.434 \mu$ and vertical intercept $\log_{10} I_0$. Using semilog paper I is plotted on the logarithmic axis and the vertical intercept on this axis is I_0.

Note the slope of the line drawn through the data points may be used to calculate μ:

$$\text{slope} = \frac{\Delta(\log_{10} I)}{\Delta x} = \frac{\log_{10} 10 - \log_{10} 100}{(3.80 - 0.40) \text{ cm}} = -.294 \text{ cm}^{-1} \tag{40}$$

From Lambert's law the theoretical slope is:

$$\text{slope} = -.434 \mu$$

Equating theoretical and experimental slopes

$$-.434 \mu = -.294 \text{ cm}^{-1}$$

and

$$\mu = +.678 \text{ cm}^{-1}$$

Plotting Data on Log-Log Paper

Log-log graph paper is used to obtain a straight line plot when y and x satisfy a power-law relation:

$$y = cx^n \tag{41}$$

where c and n are constants. For example, the semimajor axis of the orbit of a planet R is related to its period (time for one revolution around the sun) T:

$$R^3 = KT^2 \tag{42}$$

$$\text{or} \qquad R = K^{1/3} T^{2/3}$$

where K is a constant. R is nonlinearly related to T.

A straight-line plot is obtained in the following way. Taking logarithms:

$$\log_{10} R = \log_{10} (K^{1/3} T^{2/3})$$

$$= \log_{10} T^{2/3} + \log_{10} K^{1/3}$$

$$= 2/3 \log_{10} T + \log_{10} K^{1/3} \tag{43}$$

Let $y = \log_{10}R$, $x = \log_{10}T$, $m = 2/3$, and $b = \log_{10}K^{1/3}$. Then a plot of $\log_{10}R$ vs. $\log_{10}T$ would be a straight line. Log-log graph paper automatically takes the logarithm of the plotted data. A log-log graph is shown in Figure 12.

Planets: Semimajor Axis vs. Period

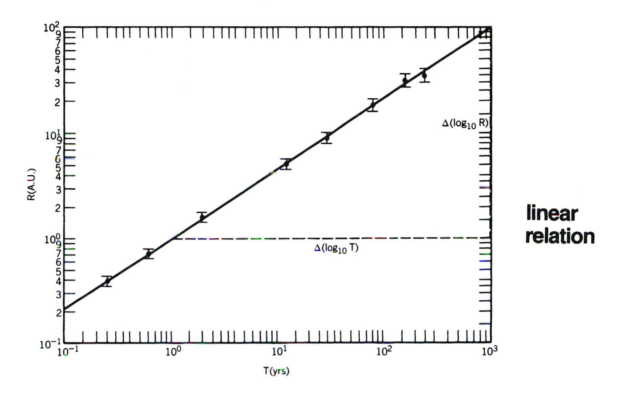

Figure 12

The units used are years and astronomical units (A.U.), where 1 A.U. is the semimajor axis of earth's orbit. (The errors shown in the graph are fictitious.) The slope of the log-log plot is:

$$\text{slope} = \frac{\Delta(\log_{10}R)}{\Delta(\log_{10}T)} = \frac{\log_{10}10^2 - \log_{10}10^0}{\log_{10}10^3 - \log_{10}10^0}$$

$$= \frac{2 - 0}{3 - 0} = \frac{2}{3} \tag{44}$$

Note that the slope of the log-log plot is the exponent of the power law relation. For example, the power law relation $y = cx^n$ plotted on log-log paper has a slope equal to n. Hence a log-log plot is a good way to determine the exponent in a power law relation.

Another way to obtain a straight-line plot is to plot y vs. x^n or R vs. $T^{2/3}$ on linear graph paper. See Figure 13.

Planets: R vs. T²/³

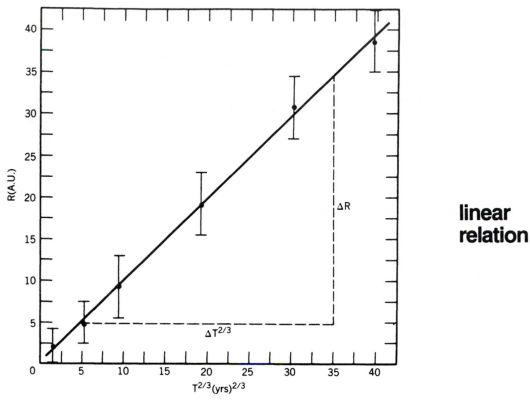

linear
relation

Figure 13

A problem with plotting R vs. $T^{2/3}$ is that values of R less than about 1 A.U. cannot be plotted with much accuracy.

In units of years and A.U. the constant K is one and inspection of the above curve shows a slope of approximately one.

Least Squares Fit

Given N data points (x_i, y_i), we would like to find the equation for the "best" straight line for this set of data. The process is sometimes called linear regression. If the measurements are distributed according to the Gauss distribution (this is usually so if the errors are random), then it can be shown that the best straight line minimizes the sum of the squares of the vertical distances d_i from each point (x_i, y_i) to the line $y = mx + b$. See Figure 14. Thus we wish to find values of m and b such that we minimize the function s, defined to be

$$s = \sum_{i=1}^{N} d_i^2 = \sum_{i=1}^{N} [y_i - (mx_i + b)]^2 \qquad (45)$$

Squaring the right-hand side, we find

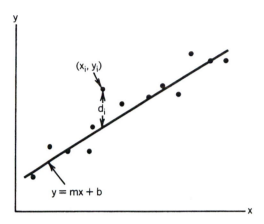

Figure 14. The best straight line minimizes the sum of the squares of the vertical distances d_i. For clarity error bars are not shown.

$$s = \Sigma(y_i)^2 - 2m\Sigma x_i y_i - 2b\Sigma y_i + m^2\Sigma x_i^2 + 2bm\Sigma x_i + Nb^2 \qquad (46)$$

where Σ is understood as a sum over the index i. Next we set

$$\frac{ds}{dm} = 0 \qquad \text{and} \qquad \frac{ds}{db} = 0 \qquad (47)$$

to find m and b corresponding to the minimum value of s. This results in two simultaneous equations:

$$\frac{ds}{dm} = -2\Sigma x_i y_i + 2m\Sigma x_i^2 + 2b\Sigma x_i = 0$$

$$\frac{ds}{db} = -2\Sigma y_i + 2m\Sigma x_i + 2Nb = 0 \qquad (48)$$

which, when solved for m and b yield

$$m = \frac{N\Sigma x_i y_i - (\Sigma x_i)(\Sigma y_i)}{N\Sigma x_i^2 - (\Sigma x_i)^2}$$

$$b = \frac{(\Sigma x_i^2)\Sigma y_i - (\Sigma x_i)(\Sigma x_i y_i)}{N\Sigma x_i^2 - (\Sigma x_i)^2} \qquad (49)$$

for the slope and intercept of the "best" line.

32

In the preceding discussion vertical distances rather than horizontal distances or some combination of vertical and horizontal distances were used. Using vertical distances in the LEAST SQUARES FIT assumes only the y_i, not the x_i, contains significant errors.

Graphical Exercises

Follow the 8 guidelines when plotting the graphs in the exercises.

(1) A cyclist starts from rest and the distance traveled is measured as a function of time. The time measurement is assumed to be precise. The data are:

d ± δd (ft)	t (s)
1.2 ± 0.5	1
5.4 ± 0.5	2
11.1 ± 0.7	3
22.0 ± 0.7	4
32.1 ± 1.0	5
49.0 ± 1.0	6
63.1 ± 1.5	7
86.0 ± 1.5	8

Determine the functional relation between distance and time by graphical analysis.

(2) Data for the world energy consumption rate are tabulated below.

P (joules/sec)	Year	t (years)
$(1.0 ± 0.5) \times 10^{12}$	1910	0
$(2.1 ± 0.5) \times 10^{12}$	1920	10
$(3.0 ± 0.5) \times 10^{12}$	1930	20
$(5.1 ± 0.5) \times 10^{12}$	1940	30
$(8.7 ± 1.0) \times 10^{12}$	1950	40
$(14.0 ± 1.5) \times 10^{12}$	1960	50
$(22.0 ± 1.5) \times 10^{12}$	1970	60

Determine the functional relation between energy consumption rate and time by graphical analysis.

RESPONSIBILITY OF THE EXPERIMENTALIST

When doing an experiment it is important for the experimentalist not to accept the data as "correct" without adequately questioning it. This is especially true if the experimentalist is not sure what to expect and/or if the data is automatically or semi-automatically recorded.

Beware of accepting data as being correct. The experimentalist should feel a sense of responsibility for the numbers (data) obtained.

It is suggested that you follow the guidelines listed below as a means of accepting the responsibility for your data.

1. If the data is semi-automatically recorded, e.g., dots on chalk tape produced by a sparker, then examine the "raw" data closely to determine whether the equipment malfunctioned.

2. After recording the data in your notebook examine the numbers for observational mistakes. Also examination of the numbers may reveal an equipment malfunction which was not previously detected.

3. Before making use of the techniques for propagating errors through the various intermediate steps and into the final result use common sense and ask yourself whether a given experimental result or error is reasonable. If it is not, there is an excellent chance either the equipment malfunctioned or you made an arithmetic or observational mistake. (Note the distinction between "mistake" and "error.")

1
DARTS AND STATISTICS

Apparatus

30-cm rule, 4 darts and a dart target for each group of 4 students. French curves for the entire lab, 3-cycle semilog paper.

Introduction

If a large number of darts are thrown at a target, is the distribution of darts describable by a normal or Gauss distribution? See NORMAL OR GAUSS DISTRIBUTION. To answer this question we do an experiment, i.e., we throw darts at a target!

Outcomes

After you have completed the activities in this experiment you will have :

a. calculated the mean dart position $\bar{x}$ and the standard deviation of the mean s_m for both your group data and the entire class data.

b. compared the distribution of dart positions to the normal or Gauss theoretical distribution using the entire class data.

Experiment

The suggested target is a sheet of 8-1/2 x 14 paper which is cut to 8-1/2 x 13 and attached to an appropriate board. See Figure 1. The target has 13 bins and each bin is 1" wide. The center bin #7, which is the bull's eye, is marked with cross hatching.   A distance of 7 or 8 feet from the dart thrower to the target is adequate.

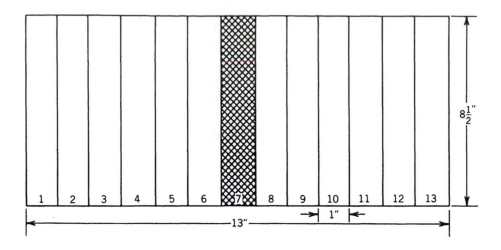

Figure 1

Each team (group of 4 students) may want to take practice throws and then attach a new target to the board.   To obtain data commence throwing darts and the more darts you throw the better your results will be.   (About 200 throws per team is a reasonable number.)   After you have finished throwing remove the target and count the number of holes in each bin.   Record your data in tabular form, letting x_i be the bin number (x_1 = 1, x_2 = 2, etc.) and $N(x_i)$ be the number of holes in the i-th bin, i = 1, 2, ..., 13. Also record your team data on the blackboard for all to see.   Table 1 shows the recommended tabular form for recording data.

Table 1

x_i	$N(x_i)$	$x_i \cdot N(x_i)$	$x_i - \bar{x}$	$(x_i - \bar{x})^2 \cdot N(x_i)$
1	$N(1)$	$1 \cdot N(1)$	$1 - \bar{x}$	$(1 - \bar{x})^2 \cdot N(x_1)$
2	$N(2)$	$2 \cdot N(2)$	$2 - \bar{x}$	$(2 - \bar{x})^2 \cdot N(x_2)$
3	$N(3)$	$3 \cdot N(3)$	$3 - \bar{x}$	$(3 - \bar{x})^2 \cdot N(x_3)$
$\vdots$	$\vdots$	$\vdots$	$\vdots$	$\vdots$
13	$N(13)$	$13 \cdot N(13)$	$13 - \bar{x}$	$(13 - \bar{x})^2 \cdot N(x_{13})$
	$\sum\limits_{i=1}^{13} N(x_i)$	$\sum\limits_{i=1}^{13} x_i \cdot N(x_i)$		$\sum\limits_{i=1}^{13} (x_i - \bar{x})^2 \cdot N(x_i)$

For this case the mean value $\bar{x}$ is given by

$$\bar{x} = \frac{\sum\limits_{i=1}^{13} N(x_i)x_i}{\sum\limits_{i=1}^{13} N(x_i)} \qquad (1)$$

where the denominator is the total number of throws n:

$$n = \sum_{i=1}^{13} N(x_i) \qquad (2)$$

The standard deviation s is:

$$s = \sqrt{\frac{1}{n-1} \sum_{i=1}^{13} (x_i - \bar{x})^2 \cdot N(x_i)} \qquad (3)$$

and the standard deviation of the mean s_m is:

$$s_m = \frac{s}{\sqrt{n}} \qquad (4)$$

See ANALYSIS OF RANDOM ERRORS. Calculate $\bar{x}$, s, and s_m using your team data and the data of the entire class.

Question 1

Calculate the ratio of the standard deviation for your team data to that for the class data. Why are the standard deviations not equal?

Plot a bar graph of $N(x_i)$ vs. x_i using the class data. A similar graph is plotted in Figure 3b of the INTRODUCTION. Do not draw the smooth curve as is done in Figure 3b. Mark the position of the mean value $\bar{x}$, obtained from the class data, on your bar graph.

Do the data follow a normal or Gauss distribution? To answer this question we compare data and theory, where the theoretical normal or Gauss distribution is given by:

$$N(x_i) = \frac{n}{\sqrt{2\pi}\sigma} e^{-(x_i-\bar{x})^2/2\sigma^2} \qquad (5)$$

In this case n is the very large number of darts thrown, and σ is the standard deviation, $\bar{x}$ is the mean, x_i is the bin number, and $N(x_i)$ is the number of dart holes in the i-th bin.

In order to calculate the theoretical number of dart holes in the i-th bin $N(x_i)$ we need $\bar{x}$, σ, and n:

1. $\bar{x}$ - We shall use the value obtained from the class data.

2. σ - Calculating the common logarithm of both sides of Equation (5) yields

$$\log_{10} N(x_i) = \log_{10} \left[\frac{n}{\sqrt{2\pi}\,\sigma} \, e^{-(x_i - \bar{x})^2 / 2\sigma^2} \right]$$

$$= \log_{10} e^{-(x_i - \bar{x})^2 / 2\sigma^2} + \log_{10} \frac{n}{\sqrt{2\pi}\,\sigma}$$

$$= -\frac{(x_i - \bar{x})^2}{2\sigma^2} \log_{10} e + \log_{10} \frac{n}{\sqrt{2\pi}\,\sigma}$$

$$= \frac{-0.434}{2\sigma^2} (x_i - \bar{x})^2 + \log_{10} \frac{n}{\sqrt{2\pi}\,\sigma} \qquad (6)$$

A plot of $\log_{10} N(x_i)$ vs. $(x_i - \bar{x})^2$ is a straight line with a theoretical slope of $-0.434/2\sigma^2$. Plot $N(x_i)$ vs. $(x_i - \bar{x})^2$ on semilog paper using the class data. See PLOTTING DATA ON SEMILOG PAPER. Measure the slope of your line and calling this number the experimental slope set it equal to the theoretical slope and solve for σ.

$$\text{experimental slope} = \text{theoretical slope}$$

$$= -\frac{0.434}{2\sigma^2} \qquad (7)$$

    3. n – The theoretical number of dart holes at the mean value $\bar{x}$ is, from Equation (5),

$$N(\bar{x}) = \frac{n}{\sqrt{2\pi}\,\sigma} \qquad (8)$$

Solve Equation (8) for n and then calculate n using σ from above and using the number of holes in the bin where $\bar{x}$ is located for $N(\bar{x})$.

    Knowing $\bar{x}$, σ, and n calculate $N(x_i)$ using Equation (5) for each bin number. Plot your points $(x_i, N(x_i))$ $i = 1, 2, \ldots, 13$ on your bar graph, plotting $N(x_i)$ in the center of the x_i interval. Draw a fine smooth curve through the points.

Question 2

    What is the percent discrepancy between σ and s? Do this calculation for s determined from your team data and the class data.

$$\text{percent discrepancy} = \frac{|\sigma - s|}{\sigma} \times 100\% \qquad (9)$$

Question 3

    Look closely at your bar graph and the smooth theoretical curve. Does the class data appear to "fit" (be describable by) the Gauss distribution curve?

2
SIMPLE PENDULUM

Apparatus

Assorted pendulum bobs, string, 2-meter stick, protractor, electric timer; one set of French curves for entire lab.

Some Definitions

A simple pendulum consists of a small mass m (the pendulum bob) suspended by a non-stretching string of length L. See Figure 1.

The period is the time for the pendulum bob to go from one extreme of position to the other and back again.

The amplitude of the pendulum's swing is the angle between the pendulum in its vertical position and at either extreme of its swing.

The length of the pendulum is the distance from the point of suspension to the center of the pendulum bob.

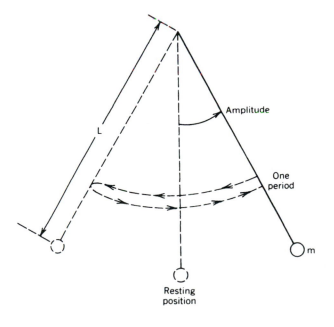

Figure 1

39

Introduction

Observation is a very common and important avenue of learning, not just in the sciences, but in many other walks of life. Often observation is supported by measurements, recording data, and representation of data by graphs. Sometimes you just observe things as they happen - whether they are being observed or not - just as in "nature." A field biologist studying grizzly bears in Yellowstone National Park would perhaps take this approach and an astronomer studying a pulsar (rotating neutron star) would definitely take this approach. Often it is necessary to deliberately manipulate conditions so you can control variables. This controlled observation is often called experimentation.

You will study the period of the simple pendulum as a function of three variables: amplitude, length, and mass. You will change only one variable at a time, and change that variable over as wide a range as possible while keeping the other two variables constant.

Outcomes

When you have made careful observations of a simple pendulum, you should be better able to:

a. minimize reaction time error.

b. use observation of several periods to get greater measurement precision than you can with measurement of a single period.

c. represent data in useful graphic form.

d. infer meaning from graphed data.

Exploration

The time interval between the input stimulus and the response to the stimulus is called the reaction time. The greatest source of error in timing the pendulum is likely to be the reaction time of the person starting and stopping the timer. The input stimulus in timing the pendulum is the visual information that the pendulum has reached a certain position and the response is pushing the electric timer button.

Question 1

To minimize the reaction time error, should the timer be started and stopped when the pendulum is at an extreme position or passing through the resting position?

Answer this question by doing an experiment! Use L = 150 cm, m = 200 gms, and amplitude = 15°. Measure the time required for the pendulum to swing through one complete period. Repeat this measurement five times for both positions of starting and stopping the timer. In both cases calculate the mean value $\bar{T}$ and the standard deviation of the mean s_m. See STATISTICAL ANALYSIS OF RANDOM ERROR.

Question 2

Which error, reaction time error or measurement error, will be more significant in this experiment? (The measurement error is likely to be the smallest division of the timer. See ESTIMATION OF RANDOM ERROR.)

The precision of the value for the pendulum period will be increased if you time n complete periods t and then calculate the time for one complete period T, where t = nT and n is an integer.

Question 3

Knowing the error in the time δt from question 2, calculate the integer n such that the fractional error in the time is less than 0.01, i.e., $\frac{\delta t}{t} < 0.01$.

Effect of Amplitude

In his book, "Dialogues Concerning Two New Sciences," Galileo stated:

> I never dreamed of learning that one and the same body, when suspended from a string a hundred cubits* long and pulled aside through an arc of 90° or even one degree or 1/2 degree would employ the same time in passing through the largest of these arcs.

Hence, Galileo concluded the period is independent of the amplitude.

Was Galileo right? Try it and see. Try 4 amplitudes from small (~ 1°) to the largest possible (~ 80°). Time 5 swings once for each amplitude. Calculate T ± δT.

Question 4

Do you find, within experimental error, the period is independent or dependent on the amplitude?

Effect of Mass

For a given (small) amplitude and a constant length L, determine T ± δT for three different masses. Again time 5 swings once for each mass.

Question 5

Does the period depend on mass within experimental error?

Effect of Length

Holding the mass and amplitude constant, determine T ± δT for five lengths ranging from about 20 cm to about 200 cm.

*1 cubit is the length from the elbow to the tip of the middle finger. (~ 18 inches)

Graphing the Data

It should be clear from the data that the period T depends on the length L. It is not clear whether T is a linear function of L or a quadratic function of L or otherwise. Often data will make more sense, and the functional relationship between variables revealed, if the data is graphed.

Following the guidelines in GRAPHICAL ANALYSIS plot T vs. L on linear graph paper.

Question 6

Is the smooth curve drawn through the data a straight line, i.e., are T and L linearly related?

Make another graph, but this time plot T vs. $L^{\frac{1}{2}}$.

Question 7

Are T and $L^{\frac{1}{2}}$ linearly related? If so, let $T = y$, $L^{\frac{1}{2}} = x$, and determine the equation of the straight line drawn through your data, i.e., determine m and b where

$$y = mx + b \qquad (1)$$

$$\text{or} \qquad T = mL^{\frac{1}{2}} + b \qquad (2)$$

where m is the slope of the line and b is the value of T where $L^{\frac{1}{2}} = 0$.

Exponent in a Power Law Relation (Optional)

Inspection of your graph for T vs. L should suggest

$$T \propto L^{n} \qquad (3)$$

where $n < 1$. Also your graph for T vs. $L^{\frac{1}{2}}$ suggests $n = \frac{1}{2}$.

Question 8

Assuming T and L obey a power law relation, a plot of T vs. L on which graph paper (semilog or log-log) should yield a straight line? See GRAPHICAL ANALYSIS.

Do such a plot, measure the slope of the line, and hence determine the exponent of the power law relation for your data.

Question 9

What is the percent discrepancy between the exponent from your graph and the theoretical exponent ($\frac{1}{2}$ or 0.5000...)?

3
ONE-DIMENSIONAL MOTION

Apparatus

Free fall apparatus with spark-timer (see Figure 1), 2-meter stick, 30-cm plastic ruler. Set of French curves for the entire lab.

Introduction

The acceleration due to Earth's gravity is an important physical constant. The U.S. Bureau of Standards reports this figure, measured at the Maryland laboratory, to be 9.810424 ± 0.000008 m/s^2. The value varies somewhat from place to place on earth, depending partly on the earth's rotation, partly on possible local concentrations of greater-than-average density (e.g., iron deposits) or less-than-average density (e.g., petroleum deposits). Of course the value also varies with the elevation above sea level.

One of the simplest physical laws is the constant acceleration of a freely falling body in a uniform gravitational field. In this experiment you will study the motion of a freely falling body and determine its constant acceleration g by two graphical methods:

1. Determination of g from plot of displacement x vs. time t.
2. Determination of g from plot of average velocity $\bar{v}$ vs. time t.

Outcomes

When you have finished the activities of this experiment, you should be able to:

a. use semi-automatically recorded data of time and position to analyze motion.

b. understand the sense of responsibility the experimentalist should feel for the numbers he or she obtains.

c. construct and interpret graphs of motion.

d. use and respect (and even admire) the value for g.

Experiment

The experiment is to be performed
using the free fall apparatus shown in Figure 1.
A metal object is initially held in place by an
electromagnet between a wire and a metal bar.
A strip of chalk coated paper is placed over
the bar. A sparker, which emits 60 sparks
per second when turned on, is connected between the bar and wire. Sparks jump from the
wire to the wide dimension of the metal object
to the metal bar, leaving a dot on the chalk-
coated paper. Thus after the metal object is
released the position of the object is recorded
every 1/60 second as a dot on the paper.

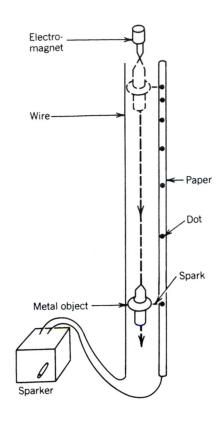

Figure 1

CAUTION: NEVER TURN ON THE
SPARKER IF ANYONE IS TOUCHING
THE FREE FALL APPARATUS.

Place the metal object so it is held by
the electromagnet. Wait until the object stops
swinging, turn on the sparker, and then turn
off the electromagnet, releasing the object.
Do not turn off the sparker until the object
hits the bottom. Remove the paper; it is your
raw data and it should remain a part of your
permanent record.

Obtain a record of metal object position vs. time as follows:

1. Tape the two ends of the paper strip to the top of the lab table. Ignoring
    the first few dots which may run into each other, number the dots as
    0, 1, 2, ..., starting from the top dot used.

    When handling data follow the three guidelines listed under RESPONSIBILITY
    OF THE EXPERIMENTALIST. From your raw data do you detect evidence
    of equipment malfunction, e.g., did the sparker fail to spark one or more
    times?

2. Using a 2-meter stick, turned on edge to eliminate parallax, measure the
    distance from dot 0 to 1, 0 to 2, 0 to 3, etc. Shift the 2-meter stick as
    few times as possible to avoid introducing new errors. Record your
    measurements and their errors in tabular form, calling the distance to
    dot 1 x_1, to dot 2 x_2, etc. Also record the time t, letting 0 seconds cor-
    respond to dot 0, 1/60 second to dot 1, 2/60 second to dot 2, etc. We
    will assume there is no error in the time per spark.

    Look closely at your recorded numbers. Are there any obvious mistakes?

Do the displacements agree with your common sense notions?

Determination of g from plot of displacement x vs. time t

Using practically a whole page for the graph, plot x vs. t from your data. Include vertical error bars on each plotted point. Carefully draw a fine smooth curve through the plotted points using French curves. See GRAPHICAL ANALYSIS. From your x vs. t graph, determine the instantaneous velocity v of the falling object at two different points. This is done by drawing lines tangent to the curve at each of the two values of t and then calculating the slopes of these tangents. Call your two results v_1 and v_2. In Figure 2 the velocity v_i at time t_i is $\Delta x_i / \Delta t_i$.

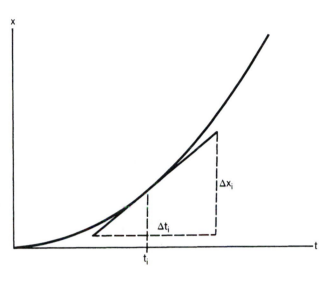

Figure 2

Question 1

What is the error in your instantaneous velocity values? (The major source of error is probably drawing the tangent line, i.e., the errors in the data are small in comparison. One way to estimate the error in the slope is to compare two independently determined values at the same point in time, i.e., compare your values of v_1 and v_2 to your partner's values. A reasonable estimation of the error is the difference between your value and your partner's value.)

The measured values v_1 and v_2 are related to the acceleration g by

$$v_1 = v_0 + gt_1 \tag{1}$$

$$v_2 = v_0 + gt_2 \tag{2}$$

where v_0 is the initial velocity. Use Equations (1) and (2) to eliminate v_0 and solve for g in terms of v_1, v_2, t_1, and t_2. Calculate g.

Question 2

What is your value of g and its error? (See PROPAGATION OF ERRORS.)

Question 3

What is the percent discrepancy between your value of g and the accepted value? (If an accepted value is not known for your local lab, use 9.80 m/s^2.)

Determination of g from plot of average velocity $\overline{v}$ vs. time t

Using your recorded data calculate the difference between the values of the displacement for successive values of the time t, i.e., $x_1 - 0$, $x_2 - x_1$, $x_3 - x_2$, $x_4 - x_3$, etc. For each difference calculate the average velocity and its error and then plot average velocity $\overline{v}$ vs. t. Your plotted points should include vertical error bars. Draw your estimate of the best straight line in the LEAST SQUARES sense. Think carefully about the instant in time which you plot average velocity against.

Question 4

What is the value of the slope and the vertical intercept of your line? Determine the error in each value. See GRAPHICAL ANALYSIS.

Question 5

What is the percent discrepancy between your value of g and the accepted value?

Timing Errors (Optional)

The precision of the timer is not perfect. The sparking circuit produces a half-wave rectified line voltage. This voltage is applied between the wire and the metal bar. When the voltage between the wire and bar is near the peak value a spark occurs. The spark may occur just before or after the voltage reaches its peak value.

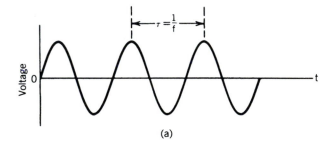

(a)

Figure 3a shows the line voltage vs. time. The period τ equals the reciprocal of the line voltage frequency f. The half-wave rectified line voltage is sketched in Figure 3b, and it has the same period τ.

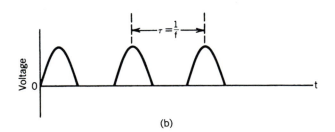

(b)

Figure 3

Question 6

Does the variation in time when the spark occurs produce a random error or a systematic error in the time per spark? See TYPES OF EXPERIMENTAL ERRORS.

In addition there may be slow variations in the 60 cycle/sec line frequency f so that it should be written as 60.0 ± 0.1 cycles/sec, say. This variation in frequency is probably slow compared to the time required for the metal object to fall, so that during the time required to record data the frequency was 59.9 cycles/sec, say.

Question 7

Does the slow variation in the line frequency cause a random or systematic error in the time per spark?

4
TWO-DIMENSIONAL MOTION

Apparatus

Ramp, steel ball, plumb bob, digital photo-timer, 30-cm rule, carbon paper, 1-meter stick, vernier calipers. Demonstration equipment: ramp, white steel ball, black background material, strobe lamp.

Introduction

This experiment involves the study of motion in two dimensions and therefore it goes beyond the study of one-dimensional motion. The most general projectile motion would involve three dimensions and could be described in terms of the x, y, and z components of the velocity and acceleration vectors. The problem is simplified using these components since the x, y, and z motions are independent of each other, i.e., the x-component of velocity is only effected by the x-component of acceleration, etc. Thus the equations describing three-dimensional motion simplify to three equations, each describing one-dimensional motion.

Projectile motion often occurs in a plane and we may orient our coordinate system such that the xy-axes lie in the plane of motion, hence there is no motion in the z direction. In addition, if one component of the acceleration in the plane is zero then we may further simplify by orienting the x-axis, say, in that direction. This is the situation in this experiment if air friction is ignored.

The projectile is a ball launched from a ramp. See Figure 1. The acceleration in the horizontal (x) direction is zero and in the vertical (y) direction is -g. In component form the position and velocity of the projectile after leaving the ramp are:

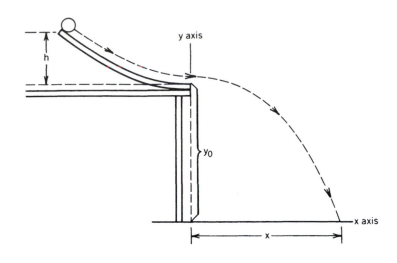

$$x = x_0 + v_{x_0} t \qquad (1)$$

$$y = y_0 + v_{y_0} t - \tfrac{1}{2} g t^2 \qquad (2)$$

$$v_x = v_{x_0} \qquad (3)$$

$$v_y = v_{y_0} - gt \qquad (4)$$

Figure 1

In the Range Experiment you will determine the speed of the ball as it leaves the ramp. Using the equations of kinematics you will calculate the range x, which may be compared with the measured value.

In the Height-Range Experiment you will measure the range x as a function of the height h. Analysis of the data will lead to a determination of the functional relationship between x and h.

Outcomes

After finishing the activities in this experiment you will have:

a. made comparison between theory and experiment.

b. used log-log graph paper to determine the functional relationship between two variables.

c. (Optional) studied the motion of an object in two dimensions and considered in detail possible causes of systematic error.

Range Experiment

Set up the photocell and lamp so that the ball will pass between them as it leaves the ramp as shown in Figure 2. (Pass a finger quickly and slowly between the lamp and photocell. Note that the timer measures the time for the finger to pass through.)

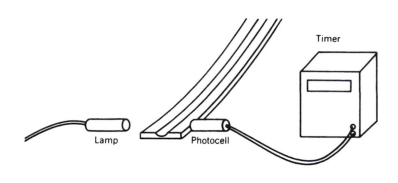

Figure 2

Release the ball near the top of the ramp and the timer will measure the time for the ball to pass the photocell. What distance does the ball move in this time? Calculate the velocity of the ball as it leaves the ramp. Use the plumb bob to locate the point on the floor just below the end of the ramp, and measure the range x from this point. To determine the point where the ball strikes the floor, tape paper there and place carbon paper - carbon side down - on the paper; the ball will print a spot on the paper.

Use equations (1) and (2) to eliminate t and solve for the range x in terms of initial condition and g. Calculate the range x.

Question 1

Do you find the measured range and the calculated range agree within experimental error?

Height-Range Experiment

What is the functional relationship between the range x and the height h? To answer this question measure x as a function of h, plot x vs. h on log-log paper, and from your graph determine x as a function of h. See PLOTTING DATA ON LOG-LOG PAPER. Draw lines of maximum and minimum slope to determine the error in the exponent of h.

In this case you are not asked to verify a physical law or to compare your experimental result with a theoretical prediction; rather you are asked to seek the answer to a question by performing a careful experiment.

Question 2

From your graph what is the functional relationship between x and h?

Perhaps it is worthwhile pointing out that in two limiting cases the theoretical relationship between x and h may be obtained. The limiting cases are:

1. the ball slides on the ramp without rolling, i.e., no friction acts between ball and ramp.

2.   the ball rolls on the ramp without sliding.

The ball probably rolls and slides on the ramp and hence the limiting cases do not represent the actual motion.

Height-Range Theory (Optional)

Assuming the ball slides without rolling, use the equations of kinematics to derive x as a function of h.

Question 3

For a given h, does the theoretical equation predict a range x which is larger than the observed value? Why is this to be expected?

Analysis of Projectile Motion (Optional)

Qualitative – Using a strobe lamp, observe a white ball against a black background as it leaves the ramp. You may find it helpful to pan the strobe lamp. If time permits repeat the observation for different time intervals per flash.

In Figure 3 the time interval between strobe flashes is constant, 0.0214 s/flash. Each dot is the location of the ball center when a flash occurred. The two marks, 20.0 ± 0.2 cm apart, were placed on the background wall to provide a scale.

Determine the average horizontal and vertical velocities, $\bar{v}_x$ and $\bar{v}_y$, between consecutive flashes. Letting the first dot in Figure 3 correspond to t = 0, plot both $\bar{v}_x$ and $\bar{v}_y$ vs. t and draw your estimates of the "best straight lines" through the data points. See LEAST SQUARES FIT. When you plot average velocity vs. time, think carefully about the point in time against which the average velocity is plotted.

Question 4

From your graphs, what are the values of a_x and a_y and their errors? Draw lines of maximum and minimum slope to estimate the error in each acceleration. See GRAPHICAL ANALYSIS.

Question 5

Do the expected accelerations, a_x = 0 and a_y = -g, and the experimental values agree within experimental errors?

Question 5

Can you conclude there was or was not a systematic error in this experiment? Briefly explain.

Print of Projectile Motion

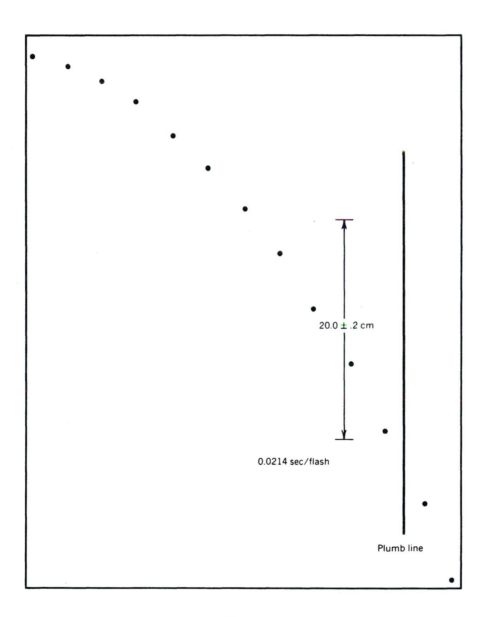

Figure 3

5

CONSERVATION OF MECHANICAL ENERGY

Apparatus

Spring, stand to support spring, set of weights, 2-meter stick, digital photo-timer, masking tape.

Introduction

Perhaps the most elegant law of physics is the conservation of mechanical energy. This law states a relationship between two distinct forms of energy: potential energy U and kinetic energy K. The relationship between K and U may be derived starting from Newton's second law:

$$F_x = \frac{mdv_x}{dt} \qquad (1)$$

Multiplying both sides by $v_x dt$ yields:

$$F_x v_x dt = mv_x dv_x \qquad (2)$$

The velocity may be written as dx/dt, the time derivative of the displacement x. Substituting dx/dt for v_x on the left-hand side of Equation (2) and then taking integrals of both sides yields:

$$\int_{x_0}^{x} F_x dx = \int_{v_0}^{v_x} mv_x dv_x \qquad (3)$$

where x_0 and v_0 are initial values of position and velocity.

Integrating the right-hand side we obtain:

$$\int_{x_0}^{x} F_x \, dx = \tfrac{1}{2} m(v_x^2 - v_0^2) \tag{4}$$

The right-hand side of Equation (4) is the change in kinetic energy of the object, which we denote by $K - K_0$. Using this notation Equation (4) becomes

$$K - K_0 = \int_{x_0}^{x} F_x \, dx \tag{5}$$

Equation (5) is the work-energy theorem, i.e., the change in kinetic energy of an object is equal to the work done on the object. If the force F_x is conservative, then the integral in Equation (5) is independent of the path from x_0 to x and it depends only on the end points, x_0 and x. Thus, for a conservative force we define

$$\int_{x_0}^{x} F_x \, dx \equiv -U(x) + U(x_0) \tag{6}$$

and the symbol U is given the name "potential energy."

Combining Equations (5) and (6), we have

$$K + U = K_0 + U_0 \tag{7}$$

where $U = U(x)$ and $U_0 = U(x_0)$. Hence if the force F_x is conservative then $K + U$, the total mechanical energy, is conserved. Some examples of conservative forces are: force of a spring, gravitational force, and the electrostatic force. Frictional forces are non-conservative.

During the past 150 years the law of energy conservation has been extended to include other forms of energy such as thermal, chemical, electrical, nuclear, radiant, etc. In its more general form the law of conservation of energy states:

> Energy cannot be created or destroyed; it may be transformed from one form to another (mechanical, thermal, electrical, nuclear, radiant, etc.), but the total energy in any isolated system never changes.

Table 1 lists examples of energy sources, the form of the energy, and the energy available.

Table 1

Source	Energy	Joules/kg
protein	chemical	1.67×10^7
gasoline	chemical	4.77×10^7
uranium	nuclear (fission)	8×10^{13}
deuterium	nuclear (fusion)	8×10^{13}
sunlight	radiant	725 Joules/m^2s*

*This is an average value at the surface of the earth. The radiation reaching the outer atmosphere of earth is 1395 Joules/m^2s; however, atmospheric absorption and reflection reduces it to the average value given in Table 1.

In this experiment you will study the mechanical energy of a mass attached to a spring. The forms of energy involved are elastic potential energy of the spring U_s, the kinetic energy of the mass K, and the mass's gravitational potential energy U_g. See Figure 1.

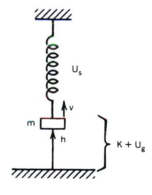

Figure 1

Outcomes

After you have completed the activities in this experiment you should have a better understanding of:

a. Hooke's law for a spring.

b. the elastic potential energy stored in a stretched spring.

c. energy conversion from elastic potential energy of a spring to kinetic energy plus gravitational potential energy.

d. conservation of mechanical energy.

Hooke's Law for a Spring

If a spring is stretched a distance x which is not too large, then Hooke's law states the spring exerts a force F which is proportional to x:

$$F = -kx \tag{8}$$

where k is the force constant of the spring and the negative sign implies F is in the opposite direction of the displacement x.

Hang the brass spring vertically, larger end down. (Record the spring number for future reference.) Place weights on the end and measure the stretch produced in each case; the stretch is just the displacement of the end of the spring from its unloaded position. See Figure 2. Loads less than 1 kg should be used; more may permanently stretch the spring. Plot the magnitude of the spring force (in Newtons) versus the stretch of the spring (in meters).

Question 1

Is your graph describable by Hooke's law? If so, determine the force constant for your spring.

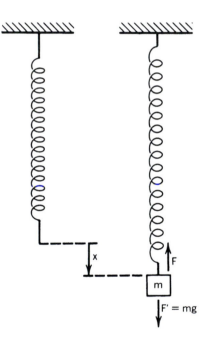

Figure 2

Elastic Potential Energy in the Spring

A load added to the spring applies a force F' which stretches the spring, storing elastic potential energy. If the added load takes the spring from some initial force F_1 and stretch x_1 to a final force F_2 and stretch x_2, then work W done by F' is given by:

$$W = \int_{x_1}^{x_2} F'dx \tag{8}$$

Assuming F' and the spring force F are equal and opposite then we may write:

$$W = \int_{x_1}^{x_2} F'\, dx = \int_{x_1}^{x_2} -F\, dx \tag{9}$$

The spring force F is conservative and from Equation (6) the right-hand side of Equation (9) equals the change in potential energy, i.e.,

$$U_s(x_2) - U_s(x_1) = -\int_{x_1}^{x_2} F\, dx = +\int_{x_1}^{x_2} kx\, dx \tag{10}$$

where the integral is just the <u>area</u> under the graph of the magnitude of the spring force versus stretch, between $(x_1, \overline{F}_1)$ and (x_2, F_2).

Question 2

From your graph what is the change in elastic potential energy of the spring when the load is increased from 0.500 kg to 0.700 kg? Give your answer in joules.

The Spring in Motion; Energy Conversions

Hang a load of 0.500 kg on the spring. Locate the equilibrium position for this load and place a photocell at this location. Hang an extra load of 0.200 kg by a thread, so that it just clears the floor. When the thread is cut, elastic potential energy in the spring is converted to kinetic energy and into gravitational potential energy as the 0.500-kg load rises.

Use the photocell to make measurements from which you can calculate the velocity of the 0.500-kg load as it passes through its equilibrium position. Attach masking tape to the 0.500-kg load and let the masking tape pass between the photocell and the lamp as shown in Figure 3.

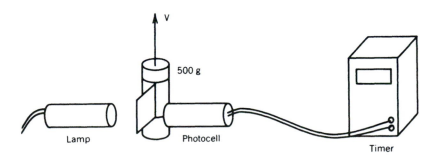

Figure 3

The initial and final energies of the 0.500-kg mass and spring are shown in Figure 4, where $U_{s,i}$ and $U_{s,f}$ are the initial and final energies of the stretched spring.

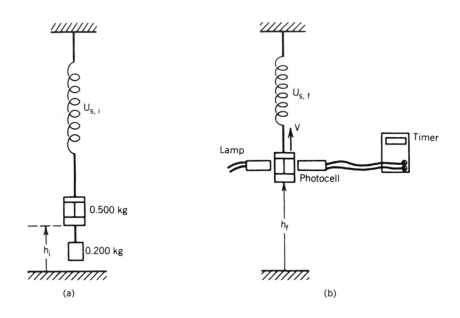

Figure 4

Equating initial and final energies

$$mgh_i + U_{s,i} = mgh_f + \tfrac{1}{2}mv^2 + U_{s,f} \qquad (11)$$

This may be rewritten as

$$U_{s,i} - U_{s,f} = mgh_f + \tfrac{1}{2}mv^2 - mgh_i \qquad (12)$$

The left side is the change in elastic potential energy of the spring and the right side is the change in energy of the 0.500-kg mass. The change in elastic potential energy is just the area under the graph of tension versus stretch which you calculated in question 2.

Question 3

Do you find that the change in elastic potential energy of the spring and the change in energy of the 0.500-kg mass agree within experimental error?

Experiment Design (Optional)

Suggest an experiment that would allow you to find a relation between your heart rate and your energy expenditure rate.

Energy Use (Optional)

Calculate the number of joules per month you use in your home and car. For the gasoline energy assume 7×10^2 kg of gasoline/m^3 and use the number of joules/kg from Table 1. You will need to estimate the number of miles per month you drive and the number of miles/gallon. Your electric bill will give electrical energy in kilowatt hours/month. If you use natural gas your bill specifies usage in therms/month, where 1 therm = 1.05×10^8 joules. If you use fuel oil the energy equivalent is 5.93×10^9 joules/barrel where one barrel is 42 gallons, and for coal the average energy equivalent is 2.95×10^7 joules/kg.

Radiant Energy (Optional)

Estimate the area of your roof. Assuming 9 hours per day of sunlight, use Table 1 to calculate the number of joules/month of radiant energy "falling" on your roof. Assuming 10% energy-conversion, i.e.,

$$0.1 = \frac{U_{out}/\text{month}}{\text{solar energy}/\text{month}}$$

calculate U_{out}/month, where U_{out}/month is the fraction of the radiant energy available per month for home use.

Would this energy, U_{out}/month, meet the energy needs of your home presently supplied by electricity, natural gas, oil and/or coal?

6

ONE-DIMENSIONAL COLLISIONS

Apparatus

Linear air track, spark-timer or two-digital photo-timers, air track cars, 1-meter stick, balance for the entire lab.

Introduction

During a collision, two objects exert internal forces, $\vec{F}_1$ and $\vec{F}_2$, on each other, which we assume satisfy Newton's third law: $\vec{F}_1 = -\vec{F}_2$. See Figure 1. If the net external force acting during the collision is zero or very small compared to the internal forces, then the change in momentum of m_1 and m_2 are:

$$\vec{P}_{1f} - \vec{P}_{1i} = \int_{t_i}^{t_f} \vec{F}_1 \, dt \qquad (1)$$

$$\vec{P}_{2f} - \vec{P}_{2i} = \int_{t_i}^{t_f} \vec{F}_2 \, dt \qquad (2)$$

where the right-hand side is the impulse delivered to the object which is the area under the internal force curve in Figure 1b and $\vec{P}_{1f} = m_1 \vec{v}_{1f}$ is the final momentum of m_1, etc.

Since we assume $\vec{F}_1 = -\vec{F}_2$, it follows from Equations (1) and (2) that

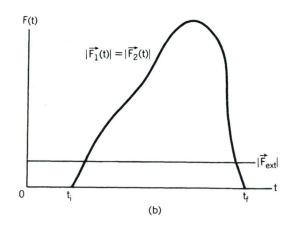

(a)

(b)

Figure 1

63

$$\vec{P}_{1f} - \vec{P}_{1i} = -(\vec{P}_{2f} - \vec{P}_{2i}) \tag{3}$$

or

$$\vec{P}_{1f} + \vec{P}_{2f} = \vec{P}_{1i} + \vec{P}_{2i} \tag{4}$$

Hence the total momentum $\vec{P}_1 + \vec{P}_2$ is not changed by the collision. In general the collision changes the momentum of both m_1 and m_2, but it does not change the total momentum.

There are three types of collisions and each type is defined by the transformation of kinetic energy that occurs.

1.   Elastic collision – An elastic collision is one that conserves kinetic energy:

$$K_{1i} + K_{2i} = K_{1f} + K_{2f} \tag{5}$$

where $K_{1i} = \frac{1}{2}m_1 v_{1i}^2$ is the initial kinetic energy of object 1, etc. The collision of two billiard balls is nearly elastic.

2.   Completely inelastic collision – A collision in which the greatest possible transformation of initial kinetic energy to other forms of energy occurs. In the laboratory frame of reference the final kinetic energy will, in general, not be zero; however, relative to the center of mass all of the kinetic energy is converted to other forms, such as heat, in the collision. In a completely inelastic collision the objects collide and stick together. An example is the collision of two balls of putty.

3.   Inelastic collision – In this case some of the initial kinetic energy is transformed to other forms. The completely inelastic collision is a special case of an inelastic collision. An example of an inelastic collision is a baseball colliding with a bat.

Important Points

a.   Momentum is conserved in all three types of collisions provided the internal forces satisfy Newton's third law and the net external force is zero or negligibly small.

b.   The total energy is conserved in all three types of collisions. In an inelastic collision some kinetic energy is converted to other forms of energy, but the total energy remains constant. In an elastic collision kinetic energy is not converted to other forms of energy.

In this experiment you will study the momentum and kinetic energy of colliding cars on an air track. One collision will be completely inelastic and the others will be nearly elastic.

Outcomes

After you have finished this experiment, you should have a better understanding of:

a. elastic and inelastic collisions.

b. conservation of momentum in both elastic and inelastic collisions.

c. conservation of kinetic energy and its role in determining if a collision is elastic or inelastic.

d. conservation of momentum which is independent of the conservation of energy, i.e., some kinetic energy is lost in inelastic collisions but the total momentum does not change.

Qualitative

Recording data is not necessary for this part.

Turn on the air supply and observe cars colliding with each other and the end of the track. Place your hand at the end of the track to feel the impulse which the colliding car delivers to the track. Arrange a variety of collisions between two cars.

Remove all cars and then place one car near the center of the track and release it. If the car accelerates to the right or left, then adjust the leveling screw until there is no acceleration. Move the car to a couple of other positions and check that the track is level.

Elastic Collision

Place a large car, with bumpers, at rest near the center of the track. Place a small car which has bumpers on the track, to the right of the large car, and give it an initial momentum directed away from the large car. (This gives the experimentalist more time before the cars collide.) Using digital photo-timers or a spark-timer, as described below, record the motion before and after the cars collide.

DIGITAL PHOTO-TIMERS

Arrange two photogates to measure the transit time of each car as shown in Figure 2.

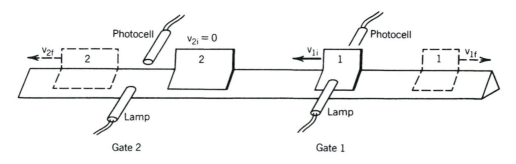

Figure 2

Note that photogate 1 is used twice: once to measure the transit time of car 1 as it moves to the left and the second time when car 1 moves to the right after colliding with car 2.

Measure the car lengths which interrupt the light beam. Choose to the left as the positive direction and calculate the velocities (magnitudes and directions).

SPARK-TIMER

Immediately after car 1 collides with the end of the air track turn on the spark-timer, recording about 30 cm of motion for car 1. Just after the cars collide turn on the spark-timer and record about 30 cm of motion. Be sure to turn off the timer before the two groups of dots corresponding to the motion of car 1 overlap.

Remove the chalk tape. Tape the ends to a flat surface, turn a meter stick on edge to eliminate parallax, and measure the three distances recorded. We will assume the time interval between dots is precisely 1/60 second. Choose to the left as positive and calculate the three velocities (magnitudes and directions).

For either method of timing, measure the mass of each car and calculate the total initial momentum and kinetic energy and the total final momentum and kinetic energy. Also determine their uncertainties.

Question 1

Within the accuracy of your measurements do you find momentum is conserved?

Question 2

There is a small external force acting during this experiment, namely the friction resulting from the air layer which supports each car. What type of error will this force produce? Is the effect of this force detectable? (Look at your answer to Question 1.)

Question 3

Is the collision elastic within the accuracy of your measurements? If not, try to account for the change in kinetic energy.

Inelastic Collision

Use two cars designed for a completely inelastic collision (they stick together). Place one at the center of the track. Collide the other car with it and record their motion using two photogates or a spark-timer.

From your data calculate the two velocities v_{1i} and v_f. Then measure the masses and calculate initial and final momentum and kinetic energy. A completely inelastic collision is shown in Figure 3.

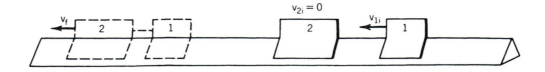

Figure 3

Question 4

Is momentum conserved within the accuracy of the experiment?

Question 5

What fraction of the initial kinetic energy is transformed to other forms?

It is possible to calculate a theoretical result for the fraction of the initial kinetic energy lost in a completely inelastic collision. If one of the cars is initially at rest, then the fraction depends only on the masses of the cars. The calculation goes as follows.

The initial kinetic energy is given by:

$$K_i = \tfrac{1}{2} m_1 v_{1i}^2 \tag{6}$$

The final kinetic energy is given by:

$$K_f = \tfrac{1}{2} (m_1 + m_2) v_f^2 \tag{7}$$

From conservation of momentum:

$$m_1 v_{1i} = (m_1 + m_2) v_f \tag{8}$$

In your lab notebook solve Equation (8) for v_f in terms of m_1, m_2, and v_{1i}. Use this result to eliminate v_f from Equation (7), then calculate the ratio of K_f to K_i. The fraction of the initial kinetic energy lost is given by:

$$\frac{K_i - K_f}{K_i} \tag{9}$$

Substituting your calculated ratio of K_f to K_i you should find:

$$\frac{K_i - K_f}{K_i} = 1 - \frac{K_f}{K_i} = \frac{m_2}{m_1 + m_2} \tag{10}$$

Question 6

What is the percent discrepancy between the theoretical fraction of kinetic energy lost and your experimental result?

7

ROTATIONAL DYNAMICS. BOOMERANG, BICYCLE

Apparatus

Rotational dynamics apparatus with two 0.5-kg masses (see Figure 6), 0.5-kg mass, 1-meter stick, vernier caliper, digital photo-timer, 2 aluminum discs (~ 8" diameter, 1/4" thick), cord, pulley. For the entire lab: three balances. Optional: ten-speed bicycle, lab jack, calipers, bathroom scales; boomerang(s) or plywood to make boomerangs.

Introduction

The rotational analog of Newton's second law may be written in two equivalent forms:

$$\vec{\tau} = I\vec{\alpha} \tag{1}$$

and

$$\vec{\tau} = \frac{d\vec{L}}{dt} \tag{2}$$

where $\vec{\tau}$ is the net torque acting on an object, I is the moment of inertia of the object, $\vec{\alpha}$ is its angular acceleration, and $\vec{L}$ is the angular momentum of the object. Using the definitions of $\vec{L}$

$$\vec{L} \equiv I\vec{\omega} \tag{3}$$

and $\vec{\alpha}$

$$\vec{\alpha} \equiv \frac{d\vec{\omega}}{dt} \tag{4}$$

where $\vec{\omega}$ is the object's angular velocity, we may readily show the equivalence of Equations (1) and (2):

$$\vec{\tau} = \frac{d\vec{L}}{dt} = \frac{d(I\vec{\omega})}{dt} = I\vec{\alpha} \tag{5}$$

Figure 1 shows three point masses mounted on an assumed massless rod and for this system

$$I = m_1 r_1^2 + m_2 r_2^2 + m_3 r_3^2 \tag{6}$$

$$|\vec{\tau}| = |\vec{r} \times \vec{F}| = rF \sin \theta \tag{7}$$

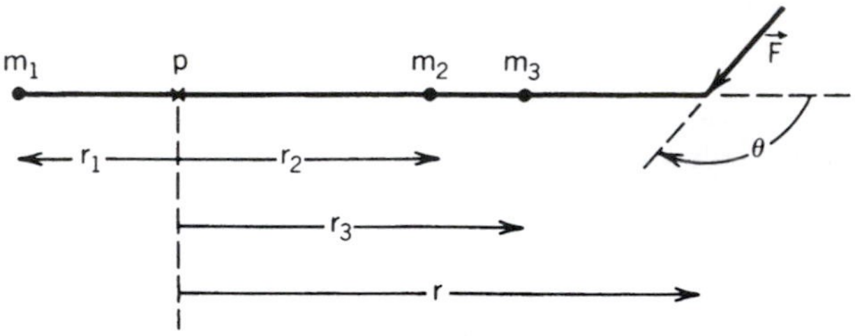

For a system composed of n point masses the moment of inertia is

$$I = \sum_{i=1}^{n} m_i r_i^2 \tag{8}$$

In the limit of a continuous mass distribution

$$\sum_i m_i r_i^2 \rightarrow \int r^2 \, dm \tag{9}$$

where r^2 is the square of the distance from the axis of rotation (we assume here planar bodies) to the small element of mass $dm = \rho \, dx \, dy$ and ρ is the density (kg/m^2).

Figure 1

A theorem about moments of inertia will be useful in doing calculations for part of this experiment. PARALLEL-AXIS THEOREM: The moment of inertia I of a body about any axis is equal to the sum of the moment of inertia about a parallel axis through the center of mass plus the total mass M of the body times the square of the distance h from the center of mass to the axis.

$$I = I_{cm} + Mh^2 \tag{10}$$

See Figure 2.

In this experiment you will study the moment of inertia of the apparatus shown in Figure 6.

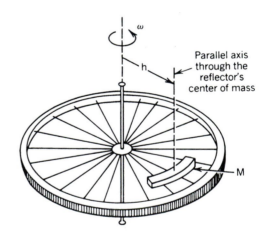

Figure 2

Outcomes

After you have finished this experiment, you should have a better understanding
of:

a.   rotational analog of Newton's second law.

b.   moments of inertia.

c.   parallel-axis theorem.

Experiment

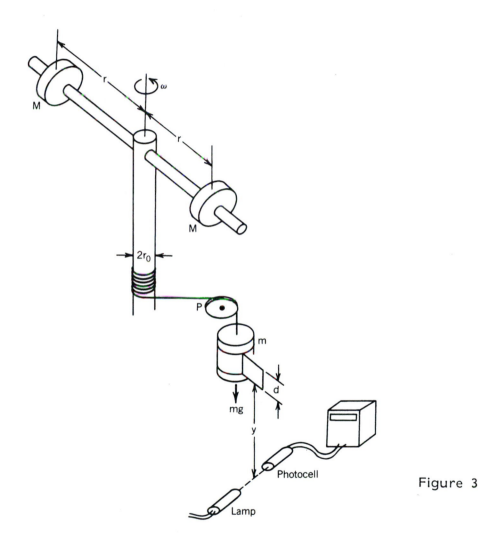

Figure 3

The rotational dynamics apparatus is shown in Figure 3. You will measure the
total moment of inertia of the apparatus shown in Figure 3 as a function of r, the distance
of each mass M from the axis of rotation. (The masses are small in size and are treated
as point masses.) The total moment of inertia I for this system is

$$I = 2Mr^2 + I_h + I_v + I_p \tag{11}$$

$$= 2Mr^2 + \text{constant}$$

where

$2Mr^2$ = moment of inertia of the two masses,

I_h = moment of inertia of the horizontal rod,

I_v = moment of inertia of the vertical rod,

I_p = effective moment of inertia of the pulley (since its ω is proportional to that of the other pieces).

How can we measure the total moment of inertia I? Well, we start by applying the rotational analog of Newton's second law to the system shown in Figure 3.

$$\tau = I\alpha \tag{12}$$

where

$$\tau = r_0 F = r_0 mg \tag{13}$$

and

$$\alpha = \frac{a}{r_0} \tag{14}$$

where a is the acceleration of the mass m. We may calculate a by releasing the mass m from rest ($v_0 = 0$) and letting it fall a known distance y. Then a is related to y and the velocity v of m at the end of the fall:

$$v^2 = 0 + 2ay \tag{15}$$

or

$$a = \frac{v^2}{2y} \tag{16}$$

To determine v, we attach a cardboard "flag" of height d to the mass m and use the digital photo-timer to measure the time t for m to fall a distance d:

$$v = \frac{d}{t} \tag{17}$$

Hence,

$$\alpha = \frac{a}{r_0} = \frac{v^2}{2r_0 y} = \frac{d^2}{2r_0 y t^2} \tag{18}$$

Substituting Equations (13) and (18) into Equation (12) and solving for I we find

$$I = \frac{2r_0^2 ymgt^2}{d^2} \qquad (19)$$

So we measure I indirectly by measuring r_0, y, m, t, and d (you measured g in the ONE-DIMENSIONAL MOTION experiment). The total moment of inertia calculated from Equation (19) is an experimentally determined value, so we define it as I_{exp}:

$$I_{exp} = \frac{2r_0^2 ymgt^2}{d^2} \qquad (20)$$

Determine I_{exp} as a function of r for five values of r. Plot I_{exp} vs. r^2.

Question 1

What is the equation of your line?

Question 2

From your graph, what is the value of $I_h + I_v + I_p$? Refer to Equation (11).

Question 3

What is the percent discrepancy between the slope of your line and the sum of the two masses, M + M? (Measure each mass with a balance.)

Parallel-Axis Theorem

Replace the two masses with the two aluminum discs. See Figure 4. (The discs are not small in size and are not treated as point masses.) In this experiment you will measure the moment of inertia of the two discs about the rotation axis shown in Figure 4, then you will use the parallel-axis theorem to calculate the moment of inertia about the center of mass of each disc. Finally you will calculate a theoretical value for the moment of inertia of a disc and compare it with your experimental result.

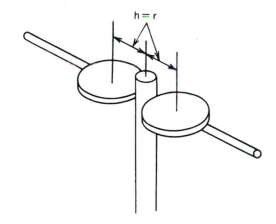

Figure 4

For the apparatus shown in Figure 4, determine I_{exp} for one value of r. Knowing the value $I_h + I_v + I_p$ from your previous results, subtract this value from I_{exp} to determine the moment of inertia of the two discs about the axis of rotation. (If time permits repeat the experiment 4 times and calculate the mean value $\bar{I}_{exp}$ and the standard deviation of the mean s_m.)

Assuming the two discs are identical use the parallel-axis theorem to determine the moment of inertia of a single disc about an axis parallel to the axis of rotation and passing through the center of mass.

The theoretical result for the moment of inertia is

$$I_{cm} = \tfrac{1}{2} MR^2 \qquad (21)$$

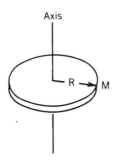

Figure 5

where M is the mass of the disc and R is the radius. See Figure 5. Measure M and R for each disc, calculate I_{cm} for each disc using Equation (21), and then calculate the mean value $\bar{I}_{cm}$.

Question 4

What is the percent discrepancy between the theoretical mean value $\bar{I}_{cm}$ and your experimental value?

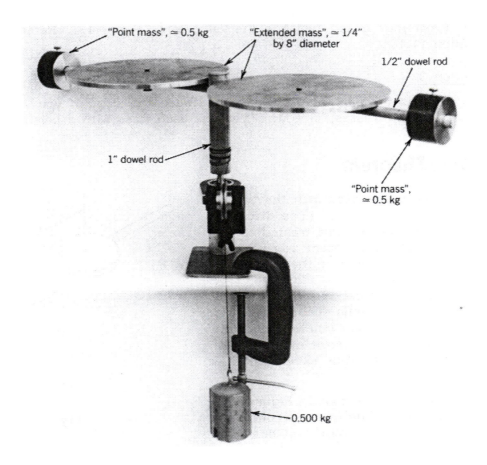

Figure 6

Boomerangs (Optional)

A two-blade boomerang for a right-handed person is shown in Figure 7. (A boomerang for a left-handed person is the mirror image of the one shown in Figure 7.) Such a boomerang may be readily made from 3/8-inch marine plywood.

To throw a boomerang grasp end "A" in your right hand and point end "B" in the direction you are throwing, with the plane of the boomerang approximately vertical. The boomerang will rotate about its center of mass with its angular momentum vector perpendicular to the plane of the boomerang as shown in Figure 8.

Throw such a boomerang and observe the change in the direction of $\vec{L}$ during the flight.

Question 5

From your observation of the change in $\vec{L}$, does the torque τ lie in the plane of the boomerang? (Ask your instructor about the forces which produce this torque.)

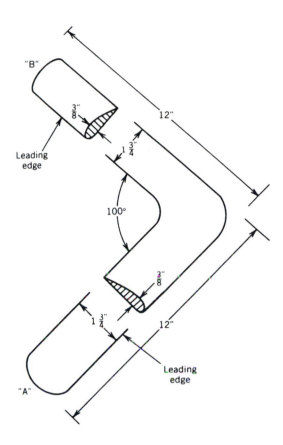

Figure 7

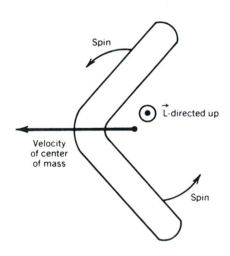

Figure 8

A four-blade boomerang is shown in Figure 9. It is interesting to compare the flight patterns of two-blade and four-blade boomerangs. Make such a comparison if a four-blade boomerang is available.

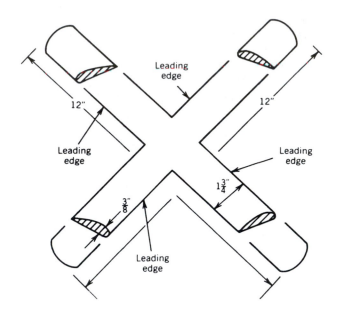

Figure 9

Bicycle (Optional)

The purpose is to calculate the hill incline, θ, which you can cycle on at constant speed, as a function of gear radii.

You need to work out the theory which relates the angle of incline, θ, to variables which we can measure in the lab. You will be guided through the theory, but expected to fill in the necessary steps.

Consider a cyclist on an incline as shown in Figure 10. Ignoring air resistance the forces acting on the system (cyclist and bike) are shown, where $\vec{F}_r$ is the force of the road on the rear wheel and $M_c\vec{g}$ and $M_B\vec{g}$ are the gravitational forces acting on the cyclist and the bike. Assume the system is moving up the incline with a constant velocity, apply Newton's second law and solve for the angle θ in terms of other variables. The other variables are measurable except for F_r, which we will eliminate as indicated below.

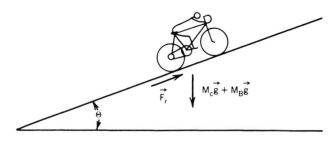

Figure 10

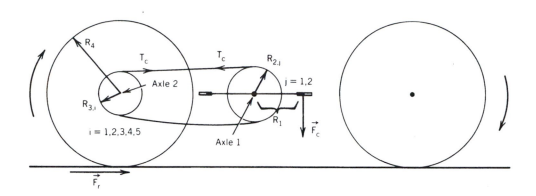

Figure 11

Using the simplified diagram of a bike shown in Figure 11, and assuming its velocity constant, apply Equation (1) to axle 1 and then to axle 2. Eliminate the chain tension T_c from the two equations and solve the resulting equation for F_r in terms of F_c (the force exerted on the pedal by the cyclist) and various radii. You may then eliminate F_r from Newton's second law and the equation gives θ as a function of measurable variables, i.e.,

$$\sin \theta = \frac{R_1 R_{3,i} F_c}{R_{2,j} R_4 (M_c + M_B) g} \tag{22}$$

Measure the various radii with calipers or a meter stick and measure the weight of cyclist and bike using the bathroom scales. Measure the force F_c by having the cyclist press down on the bathroom scales while seated on a stationary bike as shown in Figure 12. Then calculate θ for all 10 gears.

Finally, test your predictions by cycling on a nearby incline.

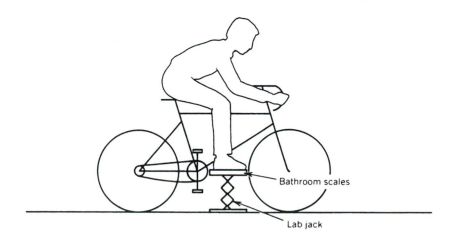

Figure 12

8
SIMPLE HARMONIC MOTION. DAMPING

Apparatus

Spring and support mechanism, weights, electric timer, 2-meter stick;
three balances for the entire lab. For the entire lab: various instruments
which exhibit damped motion. (Use the same spring as in Experiment 5.)

Introduction

When a rubber band is stretched and held extended, forces act in it which tend
to restore it to its original length. When a saw blade is bent and held, forces act in it
which tend to straighten it. In each case, the size of the elastic restoring force depends
on the dimensions of the band or blade, the amount it is stretched or bent, and the material of which it is made.

When an elastic material such as the rubber band or saw blade is strained, work
is done on it and energy is stored in it. This is a form of potential energy and most of
it can be recovered in the elastic recoil. You measured the elastic potential energy in a
spring and observed its elastic recoil producing kinetic and gravitational potential energy
in the CONSERVATION OF MECHANICAL ENERGY experiment.

If an elastic object is strained and released (or if an impulse is delivered), it will
oscillate periodically about its equilibrium or rest position. Examples of such objects are
a saw blade clamped at one end, a mass attached to a spring, a mass attached to a rod
(twisting oscillations), musical string instrument, drumhead, spider's web, eardrum, and
a car body (oscillates vertically on its springs).

Simple Harmonic Motion

When a body oscillates, if the elastic restoring force has a magnitude which is proportional to the displacement from an equilibrium position and a direction such as to restore the object to that equilibrium position, then the motion is simple harmonic. Any object in stable equilibrium will usually undergo simple harmonic oscillations about the position of stable equilibrium, if it undergoes a small initial displacement and is then re-leased.

A basic oscillator is shown in Figure 1, where k is determined by the stiffness of the elastic object (spring in this case) and the position x of the mass is measured from the point of stable equi-librium. If no other forces act, the equa-tion of motion is

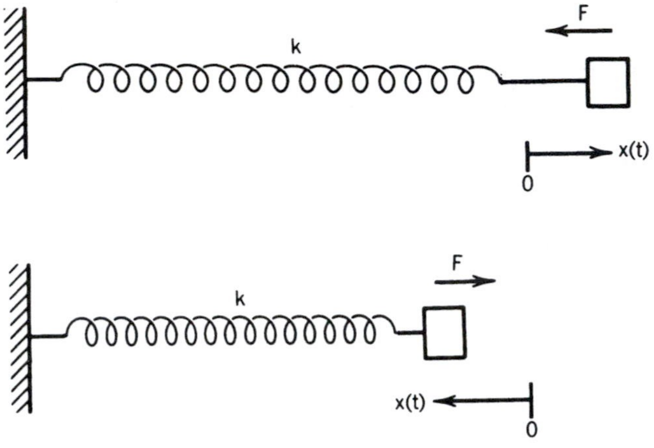

$$F = -kx = m \frac{d^2x}{dt^2} \qquad (1)$$

where in general m is the mass attached to the elastic object plus some appropriate frac-tion of the mass of the elastic object. The solution to Equation (1) is

$$x(t) = A \cos (\omega t + \phi) \qquad (2)$$

Figure 1

provided the constant angular frequency ω is

$$\omega = \sqrt{\frac{k}{m}} \qquad (3)$$

The constants A and ϕ are the amplitude and phase angle.

Equation (2) describes motion with frequency ν, the number of cycles per second,

$$\nu = \frac{\omega}{2\pi} = \frac{1}{2\pi} \sqrt{\frac{k}{m}} \qquad (4)$$

and period T, the time per cycle,

$$T = \frac{1}{\nu} = \frac{2\pi}{\omega} = 2\pi / \sqrt{\frac{k}{m}} \qquad (5)$$

The motion $x(t)$, Equation (2), is sketched in Figure 2 for arbitrary A and $\phi = 30°$.

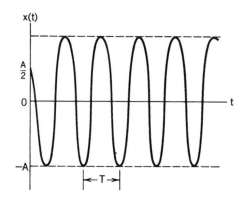

Figure 2

Question 1

Show that Equation (2) satisfies Equation (1) provided $\omega = \sqrt{k/m}$. To show this calculate d^2x/dt^2, substitute $x(t)$ and d^2x/dt^2 into Equation (1), and solve for ω.

Important point: Your work in answering Question 1 shows ω is determined by the dynamical equation of motion or Newton's second law.

Now we ask, "What determines the constants A and ϕ?" Well, A and ϕ are determined by the initial displacement x_0 and velocity v_0. To calculate A and ϕ in terms of x_0 and v_0 we proceed as follows:

$$x_0 = x(0) = A \cos (\omega t + \phi)|_{t=0} = A \cos \phi \qquad (6)$$

$$v_0 = \frac{dx}{dt}\Big|_{t=0} = -A\omega \sin (\omega t + \phi)|_{t=0} = -A\omega \sin \phi \qquad (7)$$

Squaring Equations (6) and (7), adding the squared equations and solving for A we find

$$A = \sqrt{x_0^2 + \left(\frac{v_0}{\omega}\right)^2} \qquad (8)$$

Dividing Equation (7) by (6) and solving for ϕ we find

$$\phi = \tan^{-1}\left(-\frac{v_0}{\omega x_0}\right) \qquad (9)$$

Second important point: The constants A and ϕ are determined by the initial conditions x_0 and v_0.

Question 2

If $\omega = 2$ radians/s, $x_0 = 0$, $v_0 = -0.10$ m/s, then determine $x(t)$. Sketch $x(t)$ vs. t in your lab notebook.

Damped Simple Harmonic Motion

For a real mechanical system the amplitude decreases with time and the motion is called <u>damped simple harmonic</u>. The decrease in amplitude is due to friction. The energy of oscillation is eventually dissipated as thermal energy.

The damping force is often proportional to the velocity of the mass m, but in the opposite direction. Newton's second law applied to the damped harmonic oscillator yields the equation of motion for the mass m:

$$F = -kx - b\frac{dx}{dt} = m\frac{d^2x}{dt^2} \qquad (10)$$

or

$$m\frac{d^2x}{dt^2} + b\frac{dx}{dt} + kx = 0 \qquad (11)$$

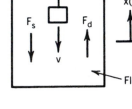

Figure 3

where b is a positive constant, called the damping constant, with units of N·s/m. A damped harmonic oscillator is shown in Figure 3. The solution to Equation (11) depends on the size of the stiffness constant k and the damping constant b. Two cases will be discussed.

(a) <u>Underdamped Simple Harmonic Motion</u>

If $\frac{k}{m} > (\frac{b}{2m})^2$, then

$$x(t) = Ae^{-bt/2m}\cos(\omega't + \phi) \qquad (12)$$

provided the constant angular frequency ω' is

$$\omega' = \sqrt{\frac{k}{m} - (\frac{b}{2m})^2} \qquad (13)$$

The period T' is

$$T' = \frac{2\pi}{\omega'} = 2\pi/\sqrt{\frac{k}{m} - (\frac{b}{2m})^2} \qquad (14)$$

and the amplitude is

$$\text{amplitude} = Ae^{-bt/2m} \qquad (15)$$

Note:

(1) Damping gives rise to an exponentially decreasing amplitude.

(2) A comparison of Equations (3) and (13) shows damping reduces the angular frequency, hence damping increases the period (compare Equations (5) and (14)).

(3) If there is no damping (b = 0), then Equation (12) reduces to Equation (2), and the motion is simple harmonic without damping.

Equation (12) is sketched in Figure 4 for ϕ = 0. This case is called <u>underdamped harmonic motion</u>. The damping constant is larger in Figure 4(b) than in Figure 4(a).

Once again the constant angular frequency ω' is determined by Newton's second law, Equation (10) or (11), i.e., if we substitute Equation (12) and its derivatives into Equation (11) and solve the resulting equation for ω', we obtain Equation (13).

The constants A and ϕ are determined by the initial conditions x_0 and v_0:

$$x_0 = x(0) = A \cos \phi \qquad (16)$$

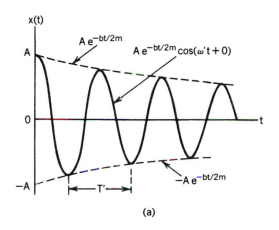

(a)

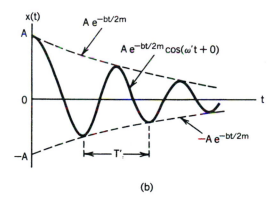

(b)

Figure 4

$$v_0 = \frac{dx}{dt}\Big|_{t=0} = -bAe^{-bt/2m} \cos(\omega't + \phi) - \omega'A^{-bt/2m} \sin(\omega't + \phi)\Big|_{t=0}$$

$$= -bA \cos \phi - \omega'A \sin \phi \qquad (17)$$

Equations (16) and (17) can be solved for A and ϕ in terms of x_0 and v_0.

(b) Overdamped Simple Harmonic Motion

If $\frac{k}{m} < (\frac{b}{2m})^2$, then the solution to Equation (11) is

$$x(t) = C_1 e^{-\gamma_1 t} + C_2 e^{-\gamma_2 t} \tag{18}$$

where γ_1 and γ_2 are determined from Newton's second law by substituting Equation (18) and its derivatives into Equation (11). The results are

$$\gamma_1 = \frac{b}{2m} + \sqrt{(\frac{b}{2m})^2 - \frac{k}{m}} \tag{19}$$

and

$$\gamma_2 = \frac{b}{2m} - \sqrt{(\frac{b}{2m})^2 - \frac{k}{m}} \tag{20}$$

The constants C_1 and C_2 are determined from the initial conditions.

Note:

    (1)   The damping constant is sufficiently large that the motion, Equation (18), is not sinusoidal in time.

    (2)   The motion, Equation (18) decreases exponentially with time.

This case is called underdamped harmonic motion and it is sketched in Figure 5 for initial conditions $x_0 \neq 0$, $v_0 = 0$. The oscillator in Figure 3 would perhaps be over-damped if the fluid was thick molasses.

In this experiment you will make qualitative observations of various instruments which exhibit damped harmonic motion, and you will study an underdamped harmonic oscillator.

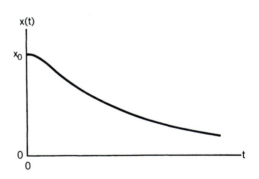

Figure 5

Outcomes

When you have finished the activities in this experiment you should:

a.   have a better understanding of simple harmonic oscillations, with and without damping.

b.   be able to analyze data plotted on semilog or log-log paper.

c. have an increased awareness (and appreciation) of the many phenomena (some man-made, some occurring naturally) which exhibit periodic motion.

Damped Systems

Many instruments which are used for measuring exhibit damped oscillations. When such instruments are used to do a measurement, a pointer or some deflection mechanism will undergo damped oscillations before settling down to the measured value. Several instruments which exhibit damped oscillations are arranged in the laboratory.

Possible instruments are:

1. Non-digital voltmeter and a 1.5-V battery
2. Bathroom scales
3. Beam balance without damping
4. Beam balance with magnetic damping

Do a measurement with each instrument and specify if the motion is underdamped or overdamped.

Have your favorite friend stand on your car bumper and step off. Observe the damping produced by the shock absorbers.

The next time you come across a spider's web, either drop a small mass into it, or pull and release it, and observe the damping.

Underdamped Simple Harmonic Oscillator

Does damping have an observable effect on the period of a mass attached to a spring and oscillating in air?

To answer this question you will:

1. determine the experimental period of the system,

2. calculate the theoretical period assuming damping is negligible, i.e.,

$$T = 2\pi\sqrt{\frac{k}{m}}$$

2. calculate the theoretical period for an underdamped harmonic oscillator,

$$T' = 2\pi\sqrt{\frac{k}{m} - \left(\frac{b}{2m}\right)^2}$$

and then you will compare the three periods.

1. EXPERIMENTAL PERIOD

Hang a medium size mass from the spring. Devise and carry out an experiment such that the fractional error in the period $\delta T_{exp}/T_{exp}$ is less than 0.01. You may want to refer to the SIMPLE PENDULUM experiment.

2. THEORETICAL PERIOD OF AN UNDAMPED SIMPLE HARMONIC OSCILLATOR

The theoretical period T is given by Equation (5):

$$T = 2\pi / \sqrt{\frac{k}{m}} \tag{21}$$

where k is the stiffness constant (determined in CONSERVATION OF MECHANICAL ENERGY experiment) and m is the effective mass (1/3 mass of the spring + attached mass). Measure the masses and calculate T and its error.

3. THEORETICAL PERIOD OF AN UNDERDAMPED SIMPLE HARMONIC OSCILLATOR

The theoretical period T' is given by Equation (14):

$$T' = 2\pi / \sqrt{\frac{k}{m} - (\frac{b}{2m})^2} \tag{22}$$

In order to calculate T' you need b, the damping constant. Determine b in the following way.

Hang the same medium size mass from the spring as used to determine the experimental period. Displace the mass from its equilibrium position by a fairly large amount but do not exceed the linear portion of the spring. Release the mass and simultaneously start the timer, then measure the amplitude and the time after every 10 complete oscillations. Measure the amplitude with a 2-meter stick standing behind the oscillating mass. Obtain 10 or more measurements and be sure to leave the timer running; hence, you will measure the amplitude as a function of time.

For an underdamped harmonic oscillator the theoretical amplitude is given by Equation (15):

$$\text{amplitude} = Ae^{-bt/2m} \tag{23}$$

Plot your data, amplitude vs. time, on the appropriate graph paper (linear, semi-log, or log-log) such that a straight line would be expected.

Question 3

Using Equation (23), what is the theoretical slope of your line? What is the numerical value of the experimental slope of your line?

Equate the theoretical and experimental slopes and then solve for b. Knowing k, m, and b, calculate the period T' using Equation (22), then calculate the error in T'.

Question 4

Which theoretical period, T or T', yields the smaller percent discrepancy when compared with the experimental period, T_{exp}?

Question 5

What is the percent discrepancy between T and T'? Is damping important with regard to the period?

9
TWO COUPLED OSCILLATORS. NORMAL MODES, RESONANCE

Apparatus

Air track, 2 small cars, 3 springs, support mechanism (for vertical suspension of springs), 2-meter stick, weights, electric timer, variable frequency driving force. For the entire laboratory: 3 balances.

Introduction

Simple harmonic oscillators were studied in the SIMPLE PENDULUM and SIMPLE HARMONIC MOTION-DAMPING experiments. In the former experiment you measured the period as a function of length for small amplitude; and in the latter experiment you studied the effects of damping.

Each oscillator, the pendulum and the mass on the spring, may be completely specified using one coordinate as shown in Figure 1. The time dependence of each coordinate is determined by solving the dynamical equation of motion (Newton's second law). Ignoring damping, the solution to the dynamical equation of motion for the mass on the spring is

$$x(t) = A \cos (\omega t + \phi) \qquad (1)$$

provided $\omega = \sqrt{k/m}$. For small amplitude, the

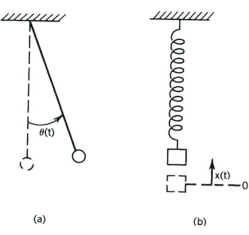

(a)   (b)

Figure 1

solution for the pendulum is

$$\theta(t) = A \cos (\omega t + \phi) \tag{2}$$

provided $\omega = \sqrt{g/\ell}$. In both cases the constants A and ϕ are determined by the initial conditions.

The number of degrees of freedom of a system is equal to the number of coordinates required to completely specify the system at any time t. Hence, each of the above oscillators has one degree of freedom. The system shown in Figure 2 where the masses oscillate horizontally, requires 2 coordinates, $x_1(t)$ and $x_2(t)$, to completely specify it, hence it has 2 degrees of freedom.

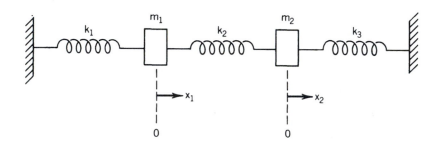

Figure 2

The number of eigenfrequencies or natural frequencies of oscillations is equal to the number of degrees of freedom. Each oscillator shown in Figure 1 has a single eigenfrequency. For the mass on the spring the eigenfrequency is

$$\omega = \sqrt{k/m} \tag{3}$$

and for the pendulum it is

$$\omega = \sqrt{g/\ell} \tag{4}$$

The system shown in Figure 2 has 2 eigenfrequencies which you will determine below.

The general motion of a system having 2 or more degrees of freedom is not simple harmonic, i.e., in general, each object (m_1 and m_2 in Figure 2) does not oscillate with a single eigenfrequency. It is possible for the motion to be simple harmonic but only if particular initial conditions are imposed, and the motion in such a case is called a normal mode of oscillation.

Consider the system shown in Figure 3 which has 2 identical springs and 2 identical masses, where x_1 and x_2 specify the displacements from equilibrium. Ignoring damping forces, Newton's second law applied to the mass on the left gives

$$F = -kx_1 - k'(x_1 - x_2) = m \frac{d^2x_1}{dt^2} \tag{5}$$

or

$$m \frac{d^2x_1}{dt^2} + (k + k')x_1 - k'x_2 = 0 \qquad (6)$$

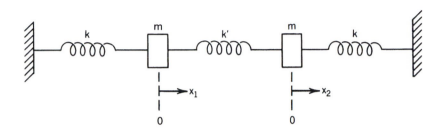

Figure 3

and for the mass on the right

$$F = -kx_2 - k'(x_2 - x_1) = m \frac{d^2x_2}{dt^2} \qquad (7)$$

or

$$m \frac{d^2x_2}{dt^2} + (k + k')x_2 - k'x_1 = 0 \qquad (8)$$

You should convince yourself that these equations are correct.

A solution to Equations (6) and (8) is

$$x_1(t) = x_2(t) = A \cos \omega t \qquad (9)$$

provided the eigenfrequency is $\omega = \sqrt{k/m}$. In this case the motion of each mass is simple harmonic and the motion is called the <u>symmetric normal mode</u> of oscillation. The masses oscillate <u>in phase,</u> with the central spring not being stretched or compressed.

Question 1

Show that Equation (9) satisfies Equation (6) or (8), provided $\omega = \sqrt{k/m}$.

Another solution to Equations (6) and (8) is

$$x_1(t) = -x_2(t) = A \cos \omega t \qquad (10)$$

provided the eigenfrequency is

$$\omega = \sqrt{\frac{k + 2k'}{m}}$$

This motion is also simple harmonic and it is called the <u>antisymmetric normal mode</u> since the two masses oscillate <u>out of phase</u>. In this case the center point of the middle spring does not move.

Question 2

Show that Equation (10) satisfies Equation (6) or (8), provided

$$\omega = \sqrt{\frac{k + 2k'}{m}}$$

The two frequencies

$$\omega = \sqrt{\frac{k}{m}} \tag{11}$$

and

$$\omega = \sqrt{\frac{k + 2k'}{m}} \tag{12}$$

are the two eigenfrequencies for the two coupled oscillators shown in Figure 3.

In this experiment you will study the harmonic and non-harmonic oscillations of two coupled oscillators. The two coupled oscillators will be two cars on air track with three springs, similar to the system in Figure 3.

Outcomes

After you finish the activities in this experiment you will have:

a. observed harmonic (normal mode) and non-harmonic oscillations of two coupled oscillators.

b. performed a single measurement for each spring in order to determine the spring constant.

c. calculated each of the two theoretical eigenfrequencies and hence the periods.

d. measured each period associated with an eigenfrequency.

Exploration

Connect two cars on the air track as shown in Figure 4 with $k_1 = k_2 = k_3$ and $m_1 = m_2$.

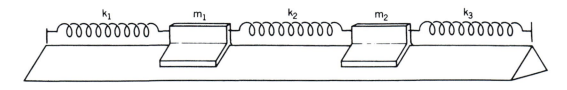

Figure 4

Use different initial conditions to set the cars in motion and observe the resulting oscillations, e.g., hold one car in its equilibrium position, displace the other car, and release both cars simultaneously. Notice that each car effects the other, i.e., the motions of each car do not remain the same for very long.

In order to excite the <u>symmetric normal mode</u>, displace both cars the same amount in the same direction, then release. Notice that the cars oscillate in phase, that each car does not affect the other, that the motion is simple harmonic, and that the middle spring does not change its length. The frequency is just what it would be if both cars were connected by a rigid rod, or if they were not connected at all.

In order to excite the <u>antisymmetric normal mode</u>, displace both cars the same amount in opposite directions, then release. Notice that the cars oscillate out of phase, that the motion of each car does not affect the other, that the motion is simple harmonic, and that the center point of the middle spring stays fixed. The frequency of the antisymmetric mode of oscillation is just what it would be if the center point of the middle spring were clamped to a vertical post, and each half treated independently.

Question 3

For the antisymmetric mode the spring constant of the middle spring is effectively twice the measured value. Why is this so?

Experimental Periods for Normal Modes

Measure the period for both the symmetric mode and the antisymmetric mode with a fractional error of less than 1%. Calculate the two periods as suggested below.

The stiffness constants of the three springs should be about the same, $k_1 \cong k_2 \cong k_3$, and the two masses should be about the same, $m_1 \cong m_2$. The average stiffness constant $\bar{k}$ is

$$\bar{k} = \frac{k_1 + k_2 + k_3}{3} \tag{13}$$

and the average mass $\bar{m}$ is

$$\bar{m} = \frac{m_1 + m_2}{2} \tag{14}$$

Using average values, the symmetric mode eigenfrequency, Equation (11), becomes

$$\omega_s = \sqrt{\frac{\bar{k}}{\bar{m}}} \tag{15}$$

and the period is

$$T_s = \frac{2\pi}{\omega_s} = 2\pi / \sqrt{\frac{\bar{k}}{\bar{m}}} \tag{16}$$

The antisymmetric mode eigenfrequency, Equation (12), becomes

$$\omega_A = \sqrt{\frac{\bar{k} + 2\bar{k}}{\bar{m}}} = \sqrt{3\frac{\bar{k}}{\bar{m}}} = \sqrt{3}\,\omega_s \qquad (17)$$

and the period is

$$T_A = \frac{2\pi}{\omega_A} = 2\pi/\sqrt{\frac{3\bar{k}}{\bar{m}}} = T_s/\sqrt{3} \qquad (18)$$

It is often useful to determine a number from a single measurement; to do so saves time and resources. Examine each spring carefully. If a given spring does not appear to be distorted from overstretching, then assume it obeys Hooke's law and determine the stiffness constant k by suspending a single mass from the spring. On the other hand, if a given spring is distorted, then study the spring in detail as you did in the CONSERVATION OF MECHANICAL ENERGY experiment.

Calculate the periods, T_A and T_s, using $\bar{k}$ and $\bar{m}$.

Question 4

Within the accuracy of your measurements do you find Equation (18) satisfied for both experimental and theoretical periods?

Question 5

Within the accuracy of your measurements do you find the theoretical and experimental results agree?

FORCED OSCILLATIONS AND RESONANCE
Introduction

In the SIMPLE HARMONIC MOTION-DAMPING experiment and so far in this experiment you have studied the natural oscillations of an object, i.e., the oscillations that occur if the object is displaced and then released. The natural frequencies for these systems are summarized in Table 1, with and without damping. For the two coupled oscillators the general motion of each mass is a superposition of the two normal mode oscillations.

Any object which is capable of natural oscillations can be made to oscillate by applying an external oscillatory force. The resulting oscillations are called forced oscillations. If the force varies sinusoidally with time at arbitrary frequency ω'', then the object will oscillate at the frequency ω'' in the steady state (after the force has been acting for a long time). The amplitude of the forced oscillations does depend on the frequency ω'' of the driving force. When ω'' is varied until the amplitude is a maximum, then the system is said to be at resonance and the value of ω'' is called the resonant frequency.

Table 1

Natural Frequencies

	Without damping	With damping
Single oscillator:	$\sqrt{\dfrac{k}{m}}$	$\sqrt{\dfrac{k}{m} - \dfrac{b^2}{4m^2}}$
Two-coupled oscillators –		
symmetric mode:	$\sqrt{\dfrac{\bar{k}}{\bar{m}}}$	$\sqrt{\dfrac{\bar{k}}{\bar{m}} - \dfrac{b^2}{4\bar{m}^2}}$
antisymmetric mode:	$\sqrt{\dfrac{3\bar{k}}{\bar{m}}}$	$\sqrt{\dfrac{3\bar{k}}{\bar{m}} - \dfrac{b^2}{4\bar{m}^2}}$

The number of resonant frequencies of a system is determined by the number of natural or eigenfrequencies. There is one resonant frequency for each natural frequency. If the damping constant b is very small, then each resonant frequency is approximately equal to a natural frequency without damping. For the single oscillator the resonant frequency is approximately $\sqrt{k/m}$ and for the two-coupled oscillators there are two resonant frequencies, approximately $\sqrt{\bar{k}/\bar{m}}$ and $\sqrt{3\bar{k}/\bar{m}}$.

Figure 5 shows a damped harmonic oscillator with an external driving force. The rotating disc provides the driving force by causing the top of the spring to oscillate with angular frequency ω'' and constant amplitude A_0. The sinusoidal motion of the top of the spring is equivalent to a sinusoidal force with angular frequency ω'' and constant amplitude F_0 acting on the system. Hence the external driving force F_D is given by

$$F_D = F_0 \cos \omega'' t \qquad (19)$$

Newton's second law for the system shown in Figure 5 is

$$F = -kx - b\frac{dx}{dt} + F_D \cos \omega'' t = m\frac{d^2x}{dt^2} \qquad (20)$$

or

$$m\frac{d^2x}{dt^2} + b\frac{dx}{dt} + kx = F_0 \cos \omega'' t \qquad (21)$$

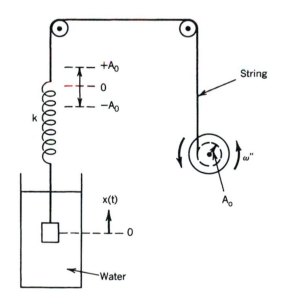

Figure 5

What is the solution, $x(t)$, to Equation (21)? Well, we expect $x(t)$ to be sinusoidal with angular frequency ω'', so we "guess"

$$x(t) = A(\omega'') \cos (\omega''t + \phi(\omega'')) \tag{22}$$

but in this case A and ϕ are not arbitrary; they are functions of the driving frequency ω''. If one substitutes $x(t)$, Equation (22), and its derivatives into Equation (21), then one finds Equation (22) is the solution to Equation (21) provided

$$\tan [\delta(\omega'')] = \frac{b\omega''}{m(\omega''^2 - \omega^2)} \tag{23}$$

$$A(\omega'') = \frac{-F_0/m}{\sqrt{(\omega''^2 - \omega^2)^2 + \frac{b^2\omega''^2}{m^2}}} \tag{24}$$

where $\omega = \sqrt{k/m}$. The maximum value of the magnitude of $A(\omega'')$ occurs for ω'' very near ω and is approximately

$$|A(\omega'')|_{max} \cong \frac{F_0}{b\omega''} \tag{25}$$

See Figure 6, where $A(\omega'')$ vs. ω'' is sketched. Note in Figure 6 that the resonant frequency is slightly less than ω for small b.

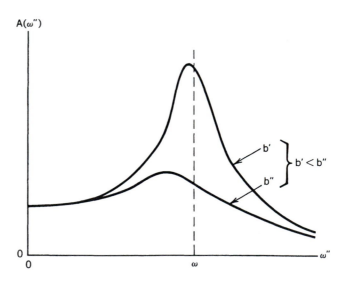

Figure 6

For two coupled oscillators each mass has two resonant frequencies, each near a normal mode frequency. The two resonant peaks are sketched in Figure 7 for either mass.

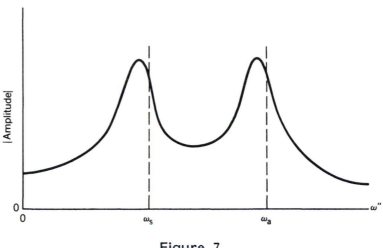

Figure 7

Experiment

Connect the driving force to the two coupled oscillators, as shown in Figure 8. (The connection of the driving force to your apparatus may differ from that shown in Figure 8.) Vary the frequency ω'' of the driving force until a resonance is reached. Use the electric timer to measure the period. Repeat for the other resonance.

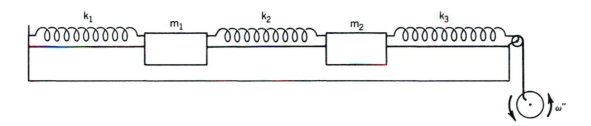

Figure 8

Question 6

What is the percent discrepancy between each natural frequency without damping and its resonant frequency?

10
OSCILLATIONS IN CONTINUOUS SYSTEMS - STRING, AIR. RESONANCE

Apparatus

String vibrator, weight holder, seven 50-gram weights, meter stick. Tube (~ 1 m long) open at one end with movable wall at the other end, small speaker, oscillator.

Introduction

VIBRATING STRING

The SIMPLE HARMONIC MOTION – DAMPING experiment is an example of an elastic system having a single discrete mass, one degree of freedom, and a single natural frequency. The TWO COUPLED OSCILLATORS – NORMAL MODES, RESONANCE experiment is an elastic system with two discrete masses, two degrees of freedom, and two natural frequencies. An elastic system with N discrete masses (where each mass moves in one dimension) has N degrees of freedom and N natural frequencies (N normal modes of oscillations).

A stretched string is an example of an elastic system that has a continuous distribution of mass. If a stretched string is appropriately plucked and released, each point on the string moves with simple harmonic motion, where the tension F in the string provides the linear restoring force. How many degrees of freedom does a vibrating string have? Well, if we divide the string of length L into N infinitesimal lengths, each of length dx and mass

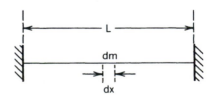

Figure 1

dm, then, assuming the string vibrates in one dimension, the system has N degrees of freedom. See Figure 1. In the limit that dx approaches zero, the stretched string has an infinite number of zero masses, hence an infinite number of degrees of freedom, an infinite number of eigenfrequencies, and an infinite number of normal modes of oscillation. Three normal modes are shown in Figure 2. The dashed line shows the displacement pattern one-half of a period later. The general equation of a normal mode is

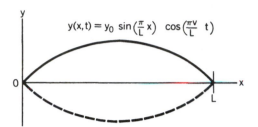

$$y(x,t) = y_0 \sin\left(\frac{n\pi}{L}x\right) \cos\left(\frac{n\pi v}{L}t\right) \quad (1)$$

where n = 1, 2, 3, ... Equation (1) is the equation for a __standing wave__ with wavelength

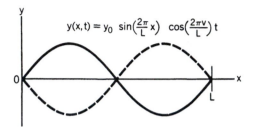

$$\lambda = \frac{2L}{n}; \quad n = 1, 2, 3, ... \quad (2)$$

and natural angular frequency

$$\omega = 2\pi v = \frac{2\pi v}{\lambda} = \frac{\pi v}{L}n \quad n = 1, 2, 3, ... \quad (3)$$

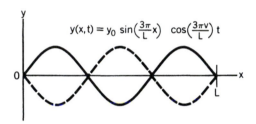

where the wave velocity v is determined by the tension F in the string and the mass per unit length μ of the string:

$$v = \sqrt{\frac{F}{\mu}} \quad (4)$$

Figure 2

Just as in the two-coupled oscillators, special initial conditions are required to excite a normal mode. The general motion of the string is a sum of normal mode oscillations.

Part of this experiment is the study of a string with one end driven sinusoidally by a fixed frequency vibrator. When the driving frequency v'' equals a natural frequency v resonance occurs and a standing wave is produced. See Figure 3.

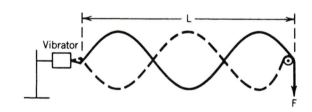

Figure 3

VIBRATING AIR IN A TUBE

A column of compressible air is like a stretched string in that it has a continuous distribution of mass, an infinite number of degrees of freedom, and an infinite number of normal modes. For this system the bulk modulus of elasticity B is analogous to the spring constant k. Three normal modes are shown in Figure 4.

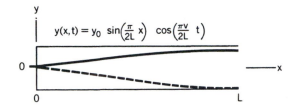

The dashed line shows the displacement one-half of a period later. The general equation for the displacement of a normal mode is

$$y = y_0 \sin\left(\frac{2\pi}{4L}(2n+1)x\right) \cos\left(\frac{2\pi(2n+1)}{4L} vt\right) \quad (5)$$

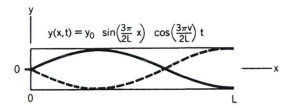

where $n = 0, 1, 2, 3, \ldots$. Equation (5) is the equation for a standing wave with wavelength

$$\lambda = \frac{4L}{2n+1} \; ; \quad n = 0, 1, 2, \ldots \quad (6)$$

and natural angular frequency

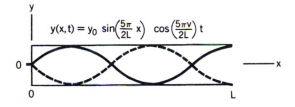

$$\omega = 2\pi\nu = \frac{2\pi v}{\lambda} = \frac{2\pi(2n+1)}{4L} v \quad (7)$$

where the velocity of sound v is determined by the bulk modulus of elasticity B and the average mass density ρ_0 of air:

Figure 4

$$v = \sqrt{\frac{B}{\rho_0}} \quad (8)$$

In this experiment you will use a small speaker connected to an oscillator as a sound source to study resonances of a tube opened at one end. See Figure 5.

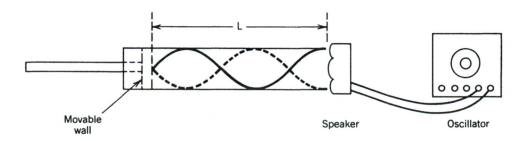

Figure 5

Outcomes

When you have finished the activities in this experiment, you will have:

a. studied standing waves on a string.
b. determined the frequency of the string vibrator.
c. studied resonances of a tube closed at one end.
d. determined the velocity of sound in air.

VIBRATING STRING

Exploration

Position the fixed frequency vibrator about 1 m from the pulley. Apply tension, F, by pulling downward on the string. Figure 3 shows a particular standing wave pattern corresponding to a certain tension. Vary the tension and observe the standing wave patterns that appear at certain tensions.

Question 1

What is the longest wavelength observed? Sketch the standing wave pattern in your notebook.

Question 2

What is the shortest wavelength observed? Sketch the pattern in your notebook.

Experiment

Combining Equations (2), (3), and (4) the natural frequency ν may be written

$$\nu = \frac{v}{\lambda} = \frac{n}{2L} \sqrt{\frac{F}{\mu}} \; ; \quad n = 1, 2, 3, \ldots \quad (9)$$

In this experiment the driving frequency ν'' is fixed, F will be determined by hanging masses on one end of the string, and L will be varied such that it always equals λ, one wavelength. See Figure 6.

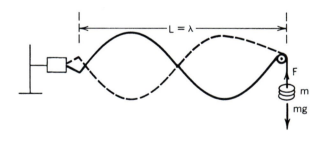

Figure 6

Attach a 50-gram weight holder to the end of the string. Slide the string vibrator along the lab table until a standing wave having maximum amplitude and one wavelength is obtained, i.e., $L = \lambda$.

Record λ and F. Add 50 grams to the weight holder and increase L until $L = \lambda$. Continue to add mass in 50-gram increments; for each mass change L so that $L = \lambda$, and record λ and F each time. Do not exceed 350 grams total. For each mass obtain a standing wave with a maximum amplitude.

Does Equation (9) describe your data? To answer this question plot some function of λ vs. F which will yield a straight line graph. From Equation (9) what function of λ should you plot vs. F in order that your graph be a straight line?

Question 3

From the slope of your line, what is the frequency of the standing wave? Your instructor will provide a value for μ.

Question 4

The manufacturer or your instructor will specify the driving frequency of the vibrator. What is the percent discrepancy between the manufacturer's value and the value from your graph?

VIBRATIONS IN AIR

Exploration

First become familiar with the speaker and oscillator by connecting the speaker leads to the output of the oscillator, then separately vary the frequency and the amplitude controls. Figure 7 is a photograph of the face of a typical oscillator.

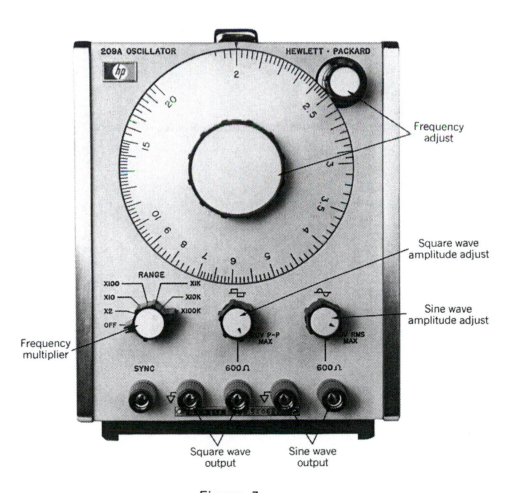

Figure 7

In this part you will use a fixed frequency sound source and vary L to produce resonances. Set the frequency of the oscillator to 500 Hz, keep the amplitude low to minimize noise pollution. Place the speaker against the tube opening as shown in Figure 5 and position the movable wall such that L = 0. Slowly increase L and listen for resonance, where the sound amplitude is a maximum.

Question 5

For each resonance observed, what is the value of n? Also, for each resonance sketch the pressure variation in the tube. See Figure 4.

Experiment

In this part you will keep L fixed and vary the driving frequency ν'' to produce resonance. Maximize L by placing the movable wall against the closed end. Increase the frequency from 50 Hz to higher values, recording 7 or 8 resonant frequencies ν_n''. Plot ν_n'' vs. n.

From Equation (7) the natural frequency is given by

$$\nu = \frac{v}{\lambda} \tag{10}$$

or

$$\nu_n = \frac{v}{\lambda_n} = \frac{v}{4L}(2n+1) \quad n = 0, 1, 2, \ldots \tag{11}$$

Question 6

From your graph determine the velocity of sound v.

11
MOTION OF FLUIDS. POISEUILLE'S LAW, DRAG FORCE

Apparatus

Fluid reservoir, 4 capillary tubes of different radii*, 200-ml beaker, 100-ml graduated cylinder, timer; 4 plastic spheres of different radii**, 100-ml beaker, 30-cm rule, micrometer or vernier caliper, glycerol.

Introduction

A. MOTION OF A FLUID THROUGH A TUBE

A fluid may flow through a tube in one of three ways: laminar flow, turbulent flow, or unstable flow. For laminar flow each particle of fluid follows a smooth regular path, which is straight if the tube is straight. Turbulent flow is agitated flow with the particles swirling about, following irregular paths. Unstable flow changes from laminar to turbulent, back to laminar, etc.

Poiseuille (pronounced Pwa-zay) developed an equation relating the flow rate $Q(m^3/s)$ of an incompressible fluid undergoing laminar flow in a horizontal tube to the

---

*Tubing 30 cm in length with inside radii of 0.25 cm, 0.20 cm, 0.15 cm, and 0.10 cm is adequate.

**Plastic spheres of nominal radii 0.75 cm, 0.60 cm, 0.45 cm, and 0.30 cm are adequate and they are available from Tap Plastic, Inc., 3011 Alvarado, San Leandro, CA 94578.

inside tube radius r(m), the fluid viscosity η (pascal sec), and the fluid pressure gradient ΔP/L (pascal/m):

$$Q = \frac{\pi r^4 \Delta P}{8 \eta L}$$ (1)

Note the strong dependence of Q on r.

The flow rate Q is inversely related to the fluid viscosity η. Viscosity is a measure of the internal (frictional) forces between adjacent layers of the fluid when these layers are in relative motion. Viscosity describes the ease with which a fluid flows. Molasses has a high viscosity and the viscosity of water is low.

Solving Equation (1) for ΔP, we obtain the pressure difference required to maintain a flow rate Q of a fluid of viscosity η in a horizontal tube of length L and radius r:

$$\Delta P = \frac{8 \eta L Q}{\pi r^4}$$ (2)

In Figure 1, $\Delta P = P_2 - P_1$. For fluids having low viscosities the pressure difference required to maintain the flow rate is less. (The viscosity of the fluid Helium-II is zero for certain low temperatures and flow speeds; it is called a <u>superfluid</u>.)

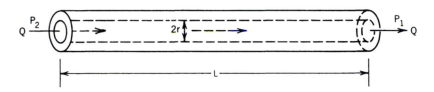

Figure 1

A theory describing the turbulent flow of a fluid through a pipe has not been developed. The solution of such problems are usually obtained by experimental analysis. An empirically determined dimensionless quantity called the <u>Reynolds number</u> N_R may be used to determine whether the flow is laminar, turbulent, or unstable. For a fluid of viscosity η, density ρ, and average speed $\bar{v}$ flowing in a tube of radius r the Reynolds number is defined to be

$$N_R = \frac{2 \rho \bar{v} r}{\eta}$$ (3)

It is found experimentally that if

$$N_R < 2000, \text{ flow is laminar}$$
$$N_R > 3000, \text{ flow is turbulent}$$
$$2000 < N_R < 3000, \text{ flow is unstable.}$$

In the first part of this experiment you will measure the flow rate Q of water as a function of tube radius r in order to determine whether the flow is laminar.

B. MOTION OF A SPHERE THROUGH A FLUID

An object moving in a fluid experiences a drag force which opposes its motion. The drag force depends on the size, shape, and speed of the object and on the density and viscosity of the fluid. Some objects are designed so that the drag is large, for example, a parachute, whereas the hull of a sailboat is designed to minimize drag. All other factors being the same, an object that is streamlined (rounded at the front and gradually tapering to a point at the back) has less drag than a non-streamlined object.

There is no simple law giving the relationship between the drag force on the object and its speed. Figure 2 shows typical experimental results: the drag is a linear function of speed at low speeds and at higher speeds it is approximately proportional to the square of the speed. v_c is the critical speed at which the drag changes from a linear to a non-linear function of speed. Streamlining an object increases v_c. At low speeds the drag force on an object is:

$$F_d = K\eta r v \qquad (4)$$

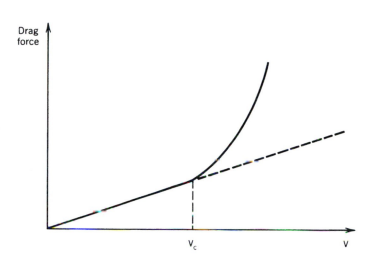

Figure 2

where K is a constant depending on the shape of the object, η is the viscosity of the fluid, r is an appropriate linear dimension, and v is the speed. Equation (4) is called the viscous or frictional drag force. (An object moving through a superfluid at low velocity does not experience a drag force.)

For a sphere of radius r, K equals 6π and the drag force is called Stoke's law:

$$F_d = 6\pi\eta r v \qquad (5)$$

At high speeds the drag force is given by

$$F_d = \frac{1}{2} C_D A \rho v^2 \qquad (6)$$

where C_D is the drag coefficient, it is a dimensionless, shape dependent constant that must be obtained from measurements. A is the cross-sectioned area (πr^2 for a sphere), ρ is the fluid density, and v is the speed of the object. The viscosity η is absent since the viscous drag force is unimportant compared with the force needed to accelerate the fluid out of the path of the object.

If Equation (4) holds, then the object passes smoothly through the fluid, leaving little turbulence in its wake. If Equation (6) holds, there will be turbulence in the wake of the object.

An empirically determined dimensionless Reynolds number may be used to determine if the drag force is best described by Equation (4) or (6). For a sphere of radius r moving with speed v in a fluid of density ρ and viscosity η, the Reynolds number N_R is

$$N_R = \frac{2\rho r v}{\eta} \qquad (7)$$

It is found experimentally that if

$$N_R < 1, \qquad F_d = 6\pi\eta r v$$

$$10^3 < N_R < 10^5, \qquad F_d = \frac{1}{2} C_D A \rho v^2$$

For intermediate values of N_R the drag force cannot be described by Equation (4) or (6), that is, it is neither a purely linear function nor a purely quadratic function of v. A general way to write the drag force so that it holds for all speeds or all values of N_R is

$$F_d = \frac{1}{2} C_D(N_R) A \rho v^2 \qquad (8)$$

where the drag coefficient $C_D(N_R)$ is now a function of N_R rather than a constant. Comparing Equation (8) with Equations (5) and (6), we find

$$C_D(N_R) = \frac{6\eta}{\frac{1}{2}r\rho v} \quad \propto \quad \frac{1}{N_R} \qquad \text{low speeds or } N_R < 1$$

$$C_D(N_R) = \text{constant} \qquad \text{high speeds or } 10^3 < N_R < 10^5$$

The drag coefficient $C_D(N_R)$ is obtained by measurements and once we know how C_D depends on N_R for an object having some particular shape, then we use Equation (8) to calculate F_d. The C_D versus N_R "master curve" for a sphere is shown in Figure 3.

In the second part of this experiment you will study spheres of different radii moving in a fluid. You will tabulate the drag coefficient $C_D(N_R)$ and the Reynolds Number N_R. You will obtain your own "master curve" by plotting C_D versus N_R.

Outcomes

After you have finished the activities in this experiment you should have a better understanding of:

a. laminar and turbulent flow.
b. fluid flow through a tube and the variables which determine the flow rate.
c. drag force which a fluid exerts on an object.
d. drag coefficient and Reynolds number.

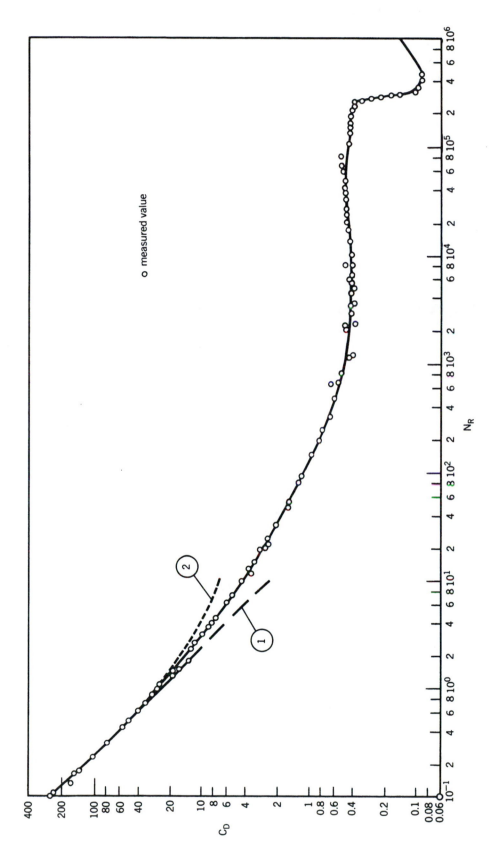

Figure 3.   Drag coefficient for spheres vs. Reynolds number.
Curve (1) is Stokes' theory.

Motion of a Fluid in a Tube

The suggested apparatus is shown in Figure 4. Record the time for approximately 100 ml of water to flow into the beaker. Calculate the flow rate Q. Repeat for each tube.

Plot Q versus r on the appropriate graph paper (regular, semi-log, or log-log) such that the slope gives the power of r for your data.

The pressure difference which drives the fluid is due to both hydrostatic and dynamic effects. The hydrostatic pressure difference ΔP_0 is given by

Figure 4

$$\Delta P_0 = \rho g h \qquad (9)$$

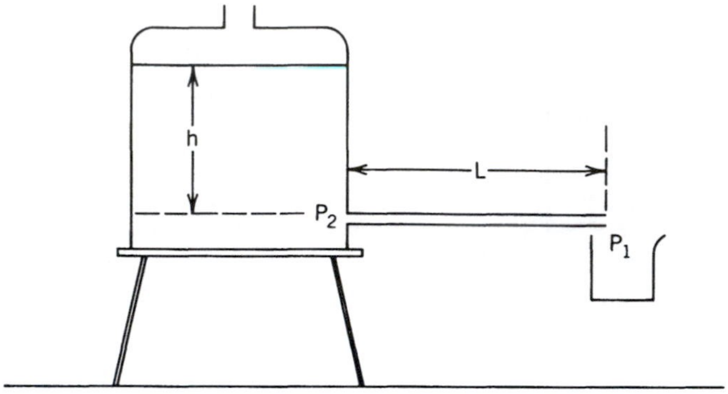

There is also a pressure drop, predicted by Bernoulli's Equation, across the entrance to the tube:

$$\Delta P_0' = \frac{1}{2}\rho v^2 \qquad (10)$$

where v is the velocity of the liquid in the tube and we assumed the velocity of the liquid in the reservoir is zero. Hence the pressure difference is

$$\Delta P \cong \rho g h - \frac{1}{2}\rho v^2 \qquad (11)$$

This may be regarded as constant for a given measurement of Q.

Question 1

What is the percent discrepancy between the slope of your line and the theoretical slope predicted by Equation (1)?

Calculate the Reynolds number for each tube using $\bar{v} = Q/\pi r^2$, $\rho = 1000 \text{ kg/m}^3$, and $\eta = 1.005 \times 10^{-3}$ Pa·s.

Question 2

For each tube does the Reynolds Number suggest the flow is laminar, turbulent, or unstable?

Motion of Spheres in a Fluid

The suggested apparatus is shown in Figure 5. Push the sphere to the bottom of the beaker, release it, and measure the time for the sphere to rise a distance y. Measure y and calculate the sphere velocity. Repeat for each sphere. Measure the diameter of each sphere using the micrometer or vernier caliper.

For each sphere calculate the Reynolds Number using Equation (7). The constants for 96% glycerol and plastic spheres are:

ρ = density of 96% glycerol = 1.25×10^3 kg/m^3

η = viscosity of 96% glycerol = 0.661 Pa·s

ρ_s = density of plastic spheres = 1.19×10^3 kg/m^3

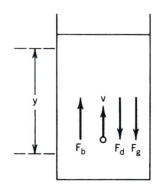

Figure 5

The density of plastic spheres is a measured value of spheres purchased from Tap Plastics, Inc.

Question 3

Does Stoke's law hold for one or more spheres?

We may obtain an expression for C_D by applying Newton's second law to the rising sphere. The forces acting on the sphere are shown in Figure 5 where:

$$F_g = \text{gravity} = mg = \frac{4}{3}\pi r^3 \rho_s g \tag{12}$$

$$F_b = \text{buoyant force} = \frac{4}{3}\pi r^3 \rho g \tag{13}$$

$$F_d = \frac{1}{2}C_D(N_R)A\rho v^2 \tag{14}$$

The sphere will quickly reach a terminal velocity; hence, the acceleration is zero and Newton's second law applied to the sphere yields:

$$0 = \Sigma F = \frac{4}{3}\pi r^3 \rho_s g - \frac{4}{3}\pi r^3 \rho g + \frac{1}{2}C_D A\rho v^2 \tag{15}$$

where downward is taken as the positive direction. Solving for C_D:

$$C_D = \frac{8}{3}rg(1 - \frac{\rho_s}{\rho})\frac{1}{v^2} \tag{16}$$

For each sphere calculate C_D.

Plot C_D versus N_R on log-log paper. Take appropriate data points from the curve in Figure 3 and plot those on the same graph with your data.

Question 4

Do you find general agreement between your data and data from the "master curve"?

Question 5

Does your graph indicate Stoke's law was satisfied for some of your data points?

12
TEMPERATURE

Apparatus

Glass bead thermistor, 1.5-V D cell, milliampere meter, thermometer, 250-ml beaker. For entire laboratory: gas filled brass cylinder fitted with a pressure gauge, ice, boiling water.

Introduction

The physical properties of a substance are temperature dependent, e.g., gas pressure, volume of a substance, electrical properties, magnetic properties, etc. The change of a physical property due to a change in temperature can be used as a thermometer.

In this experiment you will study the pressure of a gas and the electric current through a thermistor as functions of temperature.

Gases and Temperature

The particles of an ideal gas at a temperature T have a distribution of speeds which follow the Maxwell speed distribution. The number of particles, $N(v)\,dv$, having speeds between v and $v + dv$ is

$$N(v)\,dv = 4\pi N_0 \left(\frac{m}{2kT}\right)^{3/2} v^2 e^{-mv^2/2kT}\,dv$$

$$(1)$$

where T is the absolute temperature, k is the Boltzmann constant, N_0 is the total number of particles, and m is the mass of a particle. The distribution of speeds, $N(v)$ vs. v, is shown in Figure 1 for two different temperatures.

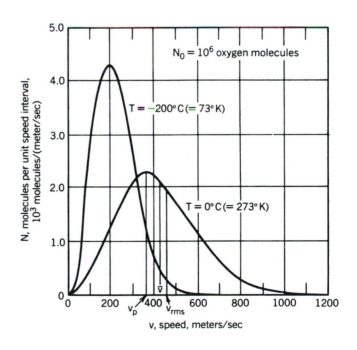

Figure 1

111

Knowing the distribution function $N(v)$, the following speeds may be readily calculated. The average speed $\bar{v}$ of the particles is

$$\bar{v} = \frac{1}{N_0} \int_0^\infty N(v) \, dv \tag{2}$$

Substituting Equation (1) for $N(v)$ and integrating:

$$\bar{v} = \left(\frac{8kT}{\pi m}\right)^{1/2} \tag{3}$$

The root-mean-square (rms) speed is

$$v_{rms} = \left(\overline{v^2}\right)^{1/2} = \left(\frac{1}{N_0} \int_0^\infty N(v)v^2 \, dv\right)^{1/2} \tag{4}$$

which, upon integrating yields

$$v_{rms} = \left(\frac{3kT}{m}\right)^{1/2} \tag{5}$$

The most probable speed v_p is the speed at which $N(v)$ has its maximum value. Requiring

$$\left.\frac{dN}{dv}\right|_{v_p} = 0 \tag{6}$$

yields

$$v_p = \left(\frac{2kT}{m}\right)^{1/2} \tag{7}$$

The average translational kinetic energy $\bar{K}$ is given by $\overline{mv^2}/2$ or from Equation (4), $\bar{K} = mv_{rms}^2/2$. Substituting Equation (5) for v_{rms}:

$$\bar{K} = \frac{3}{2} kT \tag{8}$$

Thus the absolute temperature T, a macroscopic variable, is a measure of the average translational kinetic energy of the particles, which is the average value of a microscopic variable. Note that the average translational kinetic energy of an ideal gas particle approaches zero as the absolute temperature approaches zero.

Thermistor Current and Temperature

T

Thermistors are temperature sensing devices made of semi-conducting materials. Unlike metals, for a constant voltage across the thermistor the electric current increases as the temperature increases. The temperature dependence of the current is an exponential over a limited temperature range:

$$I = I_0 e^{-\beta/T} \tag{9}$$

where β is a property of the material (expressed in degrees Kelvin) and constant over a limited temperature range.

Important point: If a system is in thermal equilibrium, then the system has a certain temperature T and the speeds or energies of the particles comprising the system are distributed according to a distribution law. For example, the distribution of speeds of an ideal gas at some temperature T obeys the Maxwell speed distribution. The distribution of the energies of the electrons in a thermistor at a temperature T is given by the Fermi-Dirac distribution law. A theoretical analysis of the thermistor current is based on this distribution law and the result of such an analysis is Equation (9).

Outcomes

After you have finished the activities in this experiment you will have:

a. determined absolute zero temperature.
b. established the Kelvin temperature scale.
c. calibrated a thermistor.

Gases and Temperature

The ideal gas law is

$$PV = nRT \tag{10}$$

where n is the number of moles, R is the Universal gas constant (8.31 J/K), and P, V, and T are the gas pressure, volume, and temperature (in degrees Kelvin). In this part of the experiment V and n remain constant and hence P varies linearly with T.

One way to do this experiment is to place the gas filled cylinder in a central location and have each student measure the gas pressure and bath temperature. Place the cylinder and a thermometer in an ice-water bath, wait about 5 minutes and then read the pressure and temperature. After each student has measured the temperature and pressure, then transfer the cylinder and thermometer to the boiling water bath, wait 5 minutes and measure temperature and pressure. The readings from the gauge is gauge pressure, not absolute pressure. To obtain absolute pressure add atmospheric pressure 1.013×10^5 N/m^2 = 14.7 lbs/inch^2) to the gauge pressure.

Plot absolute pressure vs. temperature in degrees Celsius on linear graph paper using a temperature range from -350 °C to 100 °C. As usual place error bars on the data points. Draw a line through your data points and draw lines of maximum and minimum slope. Extend all three lines so that each intersects the temperature axis.

Question 1

What is the temperature at this intersection point? Use the intersection of the lines of maximum and minimum slope to determine the error in this temperature.

Results of experiments with gases using more sophisticated equipment show this temperature to be -273.15 °C.

Question 2

Does your value of temperature agree with the above accepted value within the experimental error?

This temperature of -273.15 °C is often referred to as absolute zero. It is the lowest temperature obtainable by a system. It is convenient to define a new temperature scale such that -273.15 °C = 0° on the new scale, which is called the Kelvin scale.

Question 3

Knowing -273.15 °C = 0 K and that the degree is the same size on the Celsius and Kelvin scales, what is the general expression relating °C to K?

On your graph label the horizontal axis in K just below the Celsius scale.

Question 4

Using the cylinder and your graph, what is the temperature in this room?

Question 5

What is the rms speed of a nitrogen molecule, N_2, in this room? (The mass of a nitrogen molecule is 4.65×10^{-26} kg.)

Thermistor Current and Temperature

The suggested circuit is shown in Figure 2a. Figure 2b shows the equivalent circuit diagram. Pour boiling water into your 250-ml beaker, insert the thermistor and thermometer, and record the thermistor current and temperature. Using tap water or ice lower the bath temperature by approximately 10 °C and repeat your measurements. Continue your measurements down to approximately 0 °C.

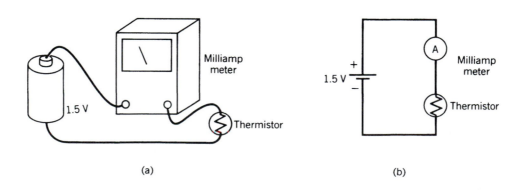

(a)   (b)

Figure 2

Over the temperature range used in this experiment, the thermistor current should obey Equation (9). Plot thermistor current vs. the appropriate function of temperature (in degrees Kelvin) on semilog paper so that a straight line is expected.

Question 6

From your graph, what are the values of β and I_0? Hence what is the experimentally determined equation which gives I as a function of T for your thermistor?

Question 7

Using your thermistor and your graph, what is the skin temperature between thumb and index finger?

13
THE OSCILLOSCOPE

Apparatus

Oscilloscope, 1.5-V D cell, oscillator, electrical leads, 6.3-V transformer (not required if the oscilloscope has a line setting on the sweep selection switch). Optional: digital photo-timer.

Introduction

The mass of the electron is very small (9.11×10^{-31} kg) so it has small inertia, i.e., a large acceleration for a small force. It is also easy to apply a force to the electron by virtue of its charge. These qualities make the electron unsurpassed as a writing instrument in the laboratory. Other instruments which record with pens and ink and paper, like strip chart recorders, are slow because the pen has a large mass. Electrons are much easier to move and a beam of electrons is used to write in TV's and oscilloscopes.

The basic component of an oscilloscope is the cathode-ray tube (CRT), shown in Figure 1. The CRT includes an electron gun, two pairs of deflection plates, and a fluorescent screen. The electron gun, which produces the beam of electrons, consists of a heater, cathode, anode, and a beam focusing mechanism that is not shown in Figure 1. If the potential difference is zero across both pairs of deflection plates, then the electron is undeflected. An undeflected beam would impact at the center of the fluorescent screen, creating a greenish spot. Atoms in the phosphor inside the screen are excited by the impacting electrons and de-excitation occurs with the emission of greenish light.

The waveform shown on the CRT screen in Figure 1 assumes that a time-varying potential difference exists across each pair of deflection plates; therefore, a time-varying electric field exists between each pair of plates. The resulting time-varying electric force on the electron beam as it passes through each pair of plates causes a vertical deflection, followed by a horizontal deflection. The electron beam spot then traces a waveform on the screen of the CRT. Because of its small mass the electron beam can follow very fast changes in potential difference (down to 10^{-9} seconds!).

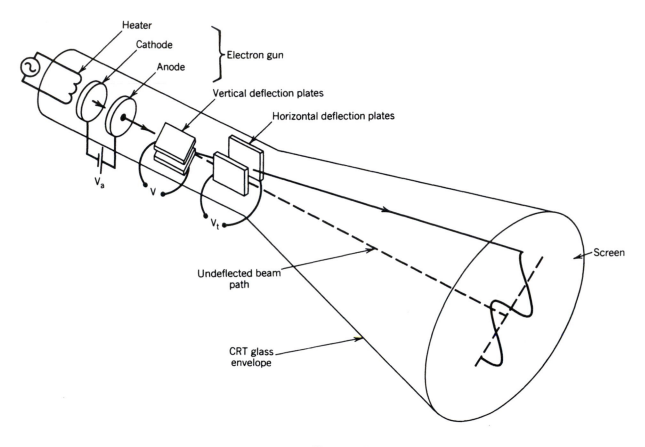

Figure 1

Figure 2 shows a block diagram of an oscilloscope.

1. The <u>power supply</u> provides the voltages necessary to produce the electron beam.

2. The <u>CRT</u> was shown in Figure 1.

3. The <u>vertical amplifier</u> amplifies the input signal to the vertical deflection plates so that small amplitude input signals produce an observable deflection of the electron beam.

4. The <u>horizontal amplifier</u> amplifies either the horizontal input signal or the sawtooth voltage which is then applied to the horizontal deflection plates.

5. The <u>sawtooth voltage generator</u> provides the sawtooth voltage that causes the electron beam spot to move from left to right at a constant (but adjustable) speed. The sawtooth voltage may be synchronized with a repetitive vertical input voltage.

We will ignore the possibility of applying a voltage to the horizontal deflection plates via the horizontal input shown in Figure 2 until part E of this experiment.

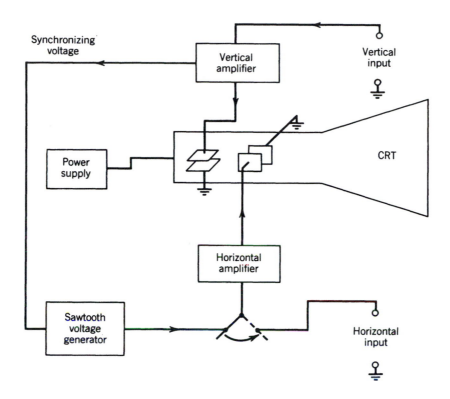

Figure 2

The sawtooth or "time-base" voltage V_t increases linearly with time as shown in Figure 3b. The period T of the sawtooth voltage is adjustable. Figure 3a shows the horizontal motion of the electron beam spot. It will be shown below that changing the period T of the sawtooth, changes the horizontal speed of the beam spot.

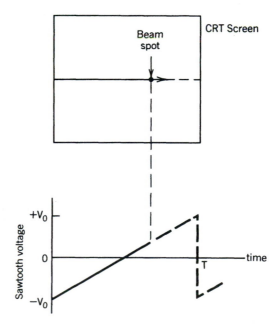

Figure 3

Electron Beam Deflection in a CRT

We derive a general expression for the angle of deflection, θ, shown in Figure 4 for some arbitrary potential V across one pair of deflection plates.

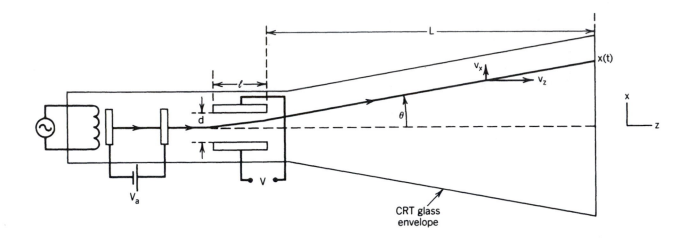

Figure 4

Electrons boiling off the hot cathode are accelerated toward the anode by an electric field which exists between the cathode and the anode. If the potential difference between cathode and anode is called V_a (the accelerating potential), then the electrons emerge from the hole in the anode with a kinetic energy

$$\frac{1}{2}mv_z^2 = eV_a \qquad (1)$$

where e is the magnitude of the electron's charge, m its mass, and the z axis is taken to be along the axis of the CRT. An electron reaches the deflection plates with

$$v_z = \sqrt{\frac{2eV_a}{m}} \qquad (2)$$

It spends a time Δt between the plates given by

$$\Delta t = \frac{\ell}{v_z} \qquad (3)$$

where ℓ is the length of the deflection plates. See Figure 4. During this time the electron experiences a force (taken to be in the x direction) given by

$$F_x = eE_x = e\frac{V}{d} \qquad (4)$$

where d is the plate separation and V is the potential between the plates. The acceleration of the electron between the plates is

$$a_x = \frac{F_x}{m} = \frac{e}{md} V \tag{5}$$

The electron has this acceleration for the time interval Δt and emerges from the plates with an x-component of velocity given by

$$v_x = a_x \Delta t = \frac{e\ell}{mdv_z} V \tag{6}$$

where Equations (3) and (5) were used. The electron then moves toward the screen at an angle θ to the axis of the CRT, with

$$\tan \theta = \frac{v_x}{v_z} = \frac{e\ell}{mdv_z^2} V \tag{7}$$

From Equation (1), we may write $mv_z^2 = 2eV_a$, hence Equation (7) becomes

$$\tan \theta = \frac{\ell}{2d} \frac{V}{V_a} \tag{8}$$

Referring to Figure 4, we may also write

$$\tan \theta = \frac{x}{L} \tag{9}$$

Combining Equations (8) and (9)

$$x = \frac{L\ell}{2dV_a} V \tag{10}$$

Important Point: ℓ, L, d, and V_a are constants, hence the displacement x of the electron beam spot is proportional to the potential V across the plates.

The vertical displacement of the electron beam spot is proportional to the potential across the vertical deflection plates and the horizontal displacement is proportional to the potential across the horizontal deflection plates.

The sawtooth or time-base voltage across the horizontal deflection plates is

$$V_t = -V_0 + \frac{2V_0}{T} t \tag{11}$$

where V_0 is a positive constant and T is the period of the sawtooth voltage. See Figure 3b. Using Equations (10) and (11) the horizontal displacement x(t) of the beam spot is

$$x(t) = \frac{L\ell}{2dV_a} \left(-V_0 + \frac{2V_0}{T} t \right) \tag{12}$$

The horizontal velocity of the beam spot is the time derivative of Equation (12)

$$\frac{dx}{dt} = \frac{L\ell}{2dV_a} \frac{2V_0}{T} \tag{13}$$

which is a constant.   Note the velocity increases as T decreases.

The principle of the oscilloscope is the following.   A time-base voltage moves the beam spot from left to right at constant speed, determined by the period T of this voltage, while an unknown voltage (which we want to study) is applied to the vertical deflection plates and swings the beam spot up and down.   Thus the oscilloscope plots voltage versus time on the CRT screen.

Typically values for the constants ℓ, d, L, and V_a are: ℓ = 2 cm, d = 0.5 cm, L = 25 cm, and V_a = 2000 V.

Question 1

Using the above constants, what is (a) the velocity v_z of an electron, (b) the time an electron spends between a pair of deflection plates, (c) the displacement x of the beam spot if the voltage at the vertical input is 2 volts and the gain (multiplier of the input voltage) of the vertical amplifier is 100, (d) the time of flight from the pair of plates to the screen (this time indicates how rapidly the electron beam responds to time-varying voltages across the plates)?

Outcomes

After you have finished this experiment, you will have:

a.   been introduced to the oscilloscope.
b.   measured DC voltage.
c.   measured the peak-to-peak voltages and the period of repetitive AC voltages.
d.   observed Lissajous patterns,
e.   measured the rise time, amplitude, and width of voltage pulses (optional).

A. Initial Oscilloscope Adjustments

The control knobs of all oscilloscopes produce basically the same effect; however, the position and labeling of the knobs does vary with the manufacturer.   The initial adjustments for the oscilloscope shown in Figure 10 are specified below.

In order to observe a DC or a time-varying voltage on the scope it is first necessary to obtain a trace (horizontal line due to the electron beam spot motion) on the screen. The procedure to establish the trace is as follows (refer to Figure 10):

1.   Turn the POWER on.

2.   Turn INTENSITY (or BRILLIANCE) knob fully clockwise.   [This makes the beam spot as bright as possible.]

3. Set the SEC/DIV (or TIME/CM) knob to 1 ms. [This sets the period T of the sawtooth voltage and hence the constant horizontal speed of the beam spot. For this setting the beam takes 1 ms to move 1 cm. The CAL (or VARIABLE) knob varies the speed of the spot between calibration. The SEC/DIV calibration is correct only when the CAL knob is in the calibrate position (pushed in or turned fully clockwise).]

4. Set the HORIZONTAL MODE to A. [This selects a mode of triggering the horizontal sweep of the beam spot. Many of the scopes used in lower-division labs will have a single horizontal mode.]

5. Set the A TRIGGER: LEVEL (or STABILITY) fully clockwise and SOURCE to INT. [The horizontal sweep of the spot is then automatically triggered by an internally supplied, (+) voltage. The LEVEL knob selects the voltage level at which the sweep is triggered.]

6. Adjust the vertical POSITION (or Y SHIFT) and the horizontal POSITION (or X SHIFT) knobs until the trace is centered on the screen.

7. Turn down the INTENSITY (or BRILLIANCE) knob to reduce the brightness. Too bright a line or dot can damage the phosphor on the screen.

8. Focus the trace by using the FOCUS knob.

The oscillosocope is now ready to be used. The scope shown in Figure 10 is a dual trace scope, i.e., it has the capability of displaying two horizontal traces and it has two vertical inputs (CH 1 and CH 2) for the simultaneous display of two waveforms. Many scopes are single trace, having one vertical input. The horizontal input is labeled EXT INPUT.

CAUTION: Make sure all "variable knobs" are in the calibrate position, otherwise your measurements will be incorrect.

B. DC Measurements

First we want to note the effect of connecting a DC voltage to the vertical input when the AC GND DC switch is set to AC. Set the switch to AC, the VOLTS/DIV to 0.5, the SEC/DIV to 1 ms, and center the trace. Connect the 1.5-V D cell to the vertical input. Observe the deflection of the trace. Reverse the leads and repeat.

Next we want to measure the voltage of the D cell. Disconnect the D cell, switch to DC, center the trace, reconnect the D cell, and measure its voltage. Reverse the leads and repeat. Switch the VOLTS/DIV to 5 and repeat.

Question 2

Does changing the VOLTS/DIV knob change the voltage input to the scope? What is changed when the VOLTS/DIV knob is changed to a different setting?

C. Peak-to-Peak Voltage Measurements

You will observe periodic voltages by connecting an oscillator to the scope. An oscillator is shown in Figure 7 of Experiment 10. The oscillator shown has two output waveforms: square wave and sine wave. For each waveform the frequency and peak-to-peak voltage are adjustable. The peak-to-peak voltage and the amplitude of a sine wave is shown in Figure 5.

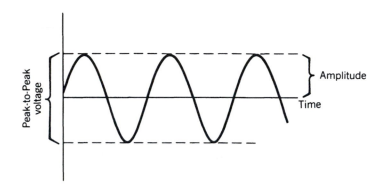

Figure 5

Leaving the initial scope adjustments as set in part A, then set the following:

VERTICAL MODE to CH 1 [For single trace scope there is only one vertical mode],

CAL in calibrate position or fully clockwise,

CH 1 VOLTS/DIV (or VOLTS/CM) to 5 (5 volts will cause a 1-cm vertical deflection],

AC GND DC switch to AC.

You should still have the trace centered on the screen.

Set the oscillator knobs initially as follows:

FREQUENCY set to 10,

FREQUENCY MULTIPLIER set to 100,

AMPLITUDE set to mid-range.

CAUTION: It is important when connecting two circuits (such as an oscilloscope and an oscillator) that their grounds are always connected together.

Connect the sine wave output of the oscillator to CH 1, vertical input, of the scope. You may have to adjust the trigger LEVEL (or STABILITY) to obtain a stable, sine wave pattern.

Change the VOLTS/DIV knob to 20 and then change it to 2, observing the wave-form on the screen for each setting.

Question 3

Does changing the VOLTS/DIV knob change the V_{pp} (peak-to-peak voltage) of the input waveform? What is changed when the VOLTS/DIV knob is moved to a different setting?

Reset the VOLTS/DIV knob to 5. Change the AMPLITUDE knob setting of the oscillator and observe the waveform on the screen.

Question 4

Does changing the AMPLITUDE knob change the V_{pp} of the waveform?

Set the AMPLITUDE knob of the oscillator so that V_{pp} is a maximum and display the waveform on the scope.

Question 5

What is the maximum value of V_{pp}?

Question 6

What VOLTS/DIV setting will result in the smallest error in V_{pp}?

If your oscillator has a square wave output, connect it, rather than the sine wave, to the scope and observe the waveform.

D. Time Measurements: Period and Frequency

Next, we want to determine the period T and the frequency ν of the waveform. To do this set the VOLTS/DIV knob to 1 and leave the SEC/DIV knob set to 1 ms. Using the sine wave output of the oscillator, set the frequency to 1000 Hz, and adjust the AMPLITUDE knob until V_{pp} is about 6 volts. If necessary adjust the trigger LEVEL (or STABILITY) knob to obtain a stable, single waveform.

Question 7

What is the period T of the waveform? To answer this question, measure the horizontal displacement corresponding to 5, say, complete periods, then calculate a single period.

Question 8

What is the percent discrepancy between your measured period and the period which you may calculate from the frequency setting of the oscillator?

Change the SEC/DIV to more than two settings, and for each setting observe the waveform. Then reset to 1 ms. Change the frequency control knobs of the oscillator and observe the effect on the waveform.

Question 9

Does changing the SEC/DIV knob change the period of the waveform? What is changed when this knob is moved to a different setting?

Question 10

Does changing the frequency control knobs of the oscillator change the period of the waveform?

E. Phase Observation: Lissajous Patterns

So far we used the oscilloscope to plot voltage versus time. The horizontal deflection was produced by a sawtooth or time-base voltage and the horizontal speed of the beam spot was constant (but adjustable).

It is often useful to apply a voltage to the terminals labeled horizontal input in Figure 2 for the horizontal deflection. On the oscilloscope, Figure 10, the voltage source connects to EXT INPUT and the A SOURCE must be switched to EXT, then the voltage is applied to the horizontal deflection plates. Switching to EXT disconnects the sawtooth voltage generator from the horizontal amplifier. See Figure 2.

Switch the oscilloscope to EXT and center the beam spot on the screen. The voltage across both pairs of deflection plates is zero and the beam is undeflected. Connect the oscillator to the horizontal deflection plates, adjust V_{pp} of the sine wave so that a 5 or 6-cm deflection of the spot occurs, then increase the frequency from the minimum to higher values, and observe the motion of the beam spot. It is simple harmonic motion. Disconnect the oscillator from the horizontal input, connect it to the vertical input and repeat the above steps. The vertical motion is also simple harmonic.

Leaving the oscillator connected to the vertical input, now connect the oscillator also to the horizontal input (see Figure 6), and adjust the oscilloscope gain control so the maximum horizontal and vertical deflections are equal. The trace on the CRT screen should be a straight line with a 45° slope. The displacements of the beam spot may be described by

$$x(t) = x_0 \cos \omega t \quad y(t) = y_0 \cos \omega t \quad (14)$$

where $x_0 = y_0$ and ω is the angular frequency, $2\pi/T$. Since the two sinusoidal signals applied to the two pairs of plates are in phase (or <u>zero phase difference</u>), then the displacements are also in phase.

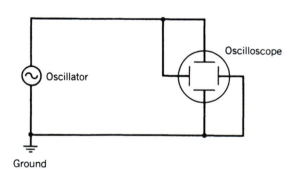

Figure 6

LISSAJOUS PATTERNS

Lissajous patterns are the superposition of two sinusoidal (simple harmonic) motions at right angles to each other. The pattern you just observed is the simplest Lissajous pattern: two motions with the same frequency, same amplitude, and same phase.

To observe other patterns, connect a 60-Hz signal to the horizontal input and the oscillator to the vertical input. Some scopes have a LINE setting on the sweep selection switch. This applies a 60-Hz, AC line signal to the horizontal input. If your scope does not have this provision, use a 6.3-V rms signal from the secondary of a transformer. Connect the sine wave of the oscillator to the vertical input. Set the frequency of the oscillator to 60 Hz, adjust the gain controls until the displacements, x_0 and y_0, are equal, then the displacements are

$$x(t) = x_0 \cos 2\pi 60t \qquad y(t) = y_0 \cos (2\pi 60t + \phi) \qquad (15)$$

where ϕ is the phase difference. The Lissajous patterns for various phase differences are shown in Figure 7. Carefully vary the frequency of the oscillator until you obtain each pattern.

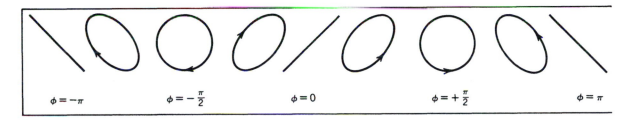

$$\phi = -\pi \qquad \phi = -\frac{\pi}{2} \qquad \phi = 0 \qquad \phi = +\frac{\pi}{2} \qquad \phi = \pi$$

Figure 7

Vary the oscillator frequency and observe more complicated patterns. In particular observe the patterns for 20, 30, 60, 120, 240, 300, and 360 Hz.

F. Voltage Pulse Measurements (Optional)

Next, we want to study a voltage pulse similar to the one shown in Figure 8. A voltage pulse is characterized by its rise time Δt_r, width Δt, and amplitude. The width is measured at half amplitude.

The photocell used in the digital photo-timer has a voltage pulse output whose width is determined by the time interval that the light beam is interrupted. We want to study the voltage pulse produced by the photocell.

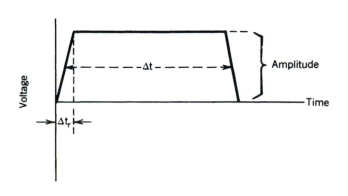

Figure 8

Connect the photocell to the vertical input of the scope as shown in Figure 9. Set the DC GND AC switch to DC, and initially set the VOLTS/DIV to 2 and the SEC/DIV to 1 ms. Using your finger to repetitively interrupt the light beam, then adjust the trigger LEVEL (or STABILITY) so that the horizontal trace triggers on the rising voltage pulse. You will probably want to change the VOLTS/DIV and SEC/DIV settings in order to adequately display the pulse.

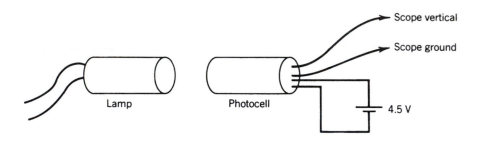

Figure 9

Next, drop an object (such as a plastic credit card) which is initially just above the light beam. Repeat, such that you may measure the rise time Δt_r, width Δt, and amplitude of the voltage pulse.

Question 11

What is the acceleration of the object? You will need to measure the dimension of the object that interrupted the light beam.

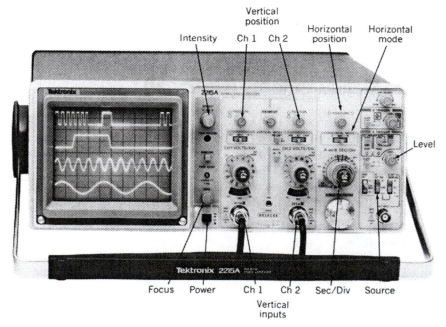

Figure 10

14

VOLTAGE, CURRENT, AND RESISTANCE MEASUREMENTS. DC AND AC

Apparatus

Oscilloscope, low voltage DC power supply, oscillator, multimeter, junction diode (1N 4000 series is adequate), 110-Ω $\frac{1}{2}$-watt resistor, 100-Ω $\frac{1}{2}$-watt resistor, electrical leads.

Introduction

For a resistive circuit element (such as a resistor, lamp bulb, or diode) the current I through the element, the voltage V across the element, and the resistance R of the element are related by

$$V = RI \tag{1}$$

If R remains constant as V and I are varied, then V is a linear function of I and the element is said to be a linear device. If R does not remain constant as V and I are varied, then V is not a linear function of I and the element is said to be a nonlinear device. A linear device satisfies Ohm's law and a nonlinear device does not.

In this experiment you will study two circuit elements, a resistor and a diode, in both an AC (alternating current) and a DC (direct current) circuit.

You will continue to use the oscilloscope and oscillator which you used in the previous experiment. An important part of this experiment is familiarization with a multimeter and its uses in measuring voltage, current, and resistance. A digital and a nondigital multimeter are shown in Figures 7 and 8. Both instruments measure voltage, current, and resistance.

Outcomes

When you have finished the activities in this experiment, you will have:

a. measured the resistance of a resistor and a diode.

b. measured the DC current through each device as a function of the potential across it and plotted the data to obtain the I-V characteristic curve.

c. displayed the AC voltage across each device on the CRT screen.

d. displayed the I-V characteristic curve for each device on the CRT screen.

A. Resistance Measurements. DC

To use your multimeter to measure resistance switch the function switch to Ω (or OHMS). The non-digital multimeter must be properly zeroed before doing the measurement. Zero the instrument by setting the resistance on the x 10 scale, short the leads (connect the ends), and adjust the knob labeled Ω (or OHMS ADJ) until the instrument reads zero ohms. Recheck the zero after changing scales and readjust if necessary.

Connect the 100-Ω carbon resistor between the probes of your multimeter and measure its resistance. Measure the resistance on the x 1, x 10, and x 100 scales. Each time you change scales recheck the zero. Reverse the probes connected to the resistor and repeat your measurements. Reversing the probes reverses the current through the resistor.

Replace the resistor with the diode and repeat the measurements, except measure the resistance of the diode on all scales.

Question 1

Which device is reversible, i.e., its resistance is independent of the direction of the current? Which device has a resistance that is dependent on the current direction?

Question 2

What is the nominal value of the resistance of the carbon resistor and its precision as specified by its color code? Does your measured value and the nominal value agree within the precision specified by the color code? The resistor color is specified in Table 1.

Table 1

Resistor Color Code

		A,B,C digits		

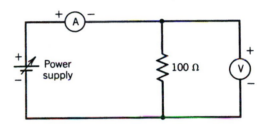

ABCD

$R = AB \times 10^C$, D

Example:
A white   B brown
C red   D silver

$R = 91 \times 10^2$, 10%

Black	0	Green	5
Brown	1	Blue	6
Red	2	Violet	7
Orange	3	Gray	8
Yellow	4	White	9

Gold (C only)   −1
Silver (C only) −2

Tolerance (D) code
Silver 10%
Gold   5%

B. Voltage and Current Measurements. DC

In order to determine whether a resistor is a linear or nonlinear electrical device you will study the resistor in an electrical circuit. Connect the circuit shown in Figure 1, using the multimeter as a DC ammeter and the oscilloscope as a DC voltmeter. (If two multimeters are available, use the second one as a DC voltmeter rather than the oscilloscope.) Turn the power supply to zero volts, then turn on the power supply and measure the current I and the potential V for three settings of the power supply. The resistor is rated at ½-watt, hence it may dissipate ½-watt or less for long periods of time. the maximum voltage across the resistor may then be calculated:

Figure 1

$$\frac{1}{2} \text{ watt} = \frac{V^2}{R} \qquad (2)$$

or

$$V = (\tfrac{1}{2} \text{ watt} \cdot R)^{\frac{1}{2}} = (50 \text{ watt} \cdot \text{ohms})^{\frac{1}{2}} = 7.1 \text{ volts} \qquad (3)$$

Do not exceed 7.1 V.

Reverse the current through the resistor by disconnecting it, rotating it 180°, and reconnecting it to the circuit; then repeat your measurements. Label the reversed current and voltage values as negative. Plot the current I through the resistor versus the voltage V across it. Place the origin at the center of the graph paper in order to plot positive and negative values.

Question 3

Does your graph suggest that the resistor is a linear device? If so, determine its resistance from the graph.

To study a semiconductor diode connect the circuit shown in Figure 2. The arrow or band on the diode indicates the direction of easy current flow, usually called the <u>forward direction</u>.

<u>CAUTION</u>: Obtain the maximum current rating of the diode from your instructor and do not exceed it; to do so would damage the diode.

You will probably want more than six data points for the diode. Plot your data, I versus V, and answer Question 3 for the diode.

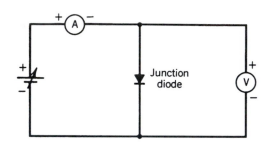

Figure 2

C. Voltage Measurements. AC

In the first part you will study a 100-Ω resistor in an AC circuit. You will:

1. Measure the AC voltage across the resistor with a multimeter and an oscilloscope, and compare them.

2. Display the characteristic I-V curve of a resistor on the scope screen.

3. Measure the phase angle between the current I through the resistor and the voltage V across it.

Connect the circuit shown in Figure 3. Note that the oscillator is "floating," that is, neither lead connects directly to ground; check your oscillator on this point. Then set the oscillator frequency to 1000 Hz and the amplitude to about half maximum. The voltage across the resistor may be described by

$$V(t) = V_0 \sin 2\pi\nu t \qquad (4)$$

where V_0 is the amplitude and the frequency is ν. Display the voltage versus time waveform on the CRT screen. The multimeter, used as an AC voltmeter, measures an average voltage, called the root-mean-square or rms voltage, V_{rms}. V_{rms} is defined to be

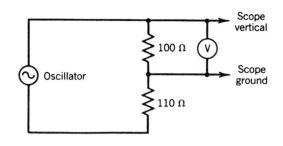

Figure 3

$$V_{rms} = (\frac{1}{T} \int_0^T V^2(t) \ dt)^{\frac{1}{2}} \tag{5}$$

where $T = 1/\nu$ is the period. For the time varying voltage $V(t)$ given by Equation (4), V_{rms} is

$$V_{rms} = (\frac{1}{T} \int_0^T V_0^2 \sin^2 2\pi\nu t \ dt)^{\frac{1}{2}} \tag{6}$$

$$= \frac{V_0}{\sqrt{2}}$$

The 120-volt ordinary household voltage is an rms value and the amplitude is $\sqrt{2} \cdot 120$ or 170 volts. (Peak-to-peak is 340 volts.) Measure V_{pp} with the scope and V_{rms} with the multimeter. Use your V_{pp} data to calculate V_{rms}.

Question 4

What is the percent discrepancy between your measured and calculated values of V_{rms}?

In part B you plotted I versus V for the resistor and diode. Such a curve is called the I-V characteristic curve for the device. To display the I-V characteristic curve for the resistor, connect the circuit shown in Figure 4. With the oscillator set at 1000 Hz, vary the oscillator amplitude and the vertical gain on the scope until the horizontal and vertical displacements of the beam spot on the CRT screen are about equal. Note that the horizontal sweep is provided by the voltage across the 100-Ω resistor. The current I(t) is proportional to the voltage across either resistor, hence the voltage applied to the vertical deflection plates is proportional to I(t). Ignoring the

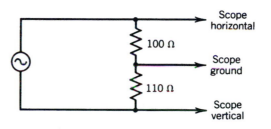

Figure 4

constant of proportionality, the CRT screen will display I versus V. Note that the common point between the resistors is grounded; thus if the voltage across the 100-Ω is positive relative to ground, then the voltage across the 110-Ω will be negative relative to ground, that is, the ground point creates a relative phase shift of 180°. Obtain the curve on the CRT screen and sketch it in your lab notebook.

Question 5

From the observed trace what is the phase difference between I(t) and V(t)? (See Figure 7, Experiment 13. Do not ignore the relative phase shift of 180° due to grounding.)

The diode in the circuit shown in Figure 5 controls the resistor current. Connect the circuit. Using the sawtooth voltage to provide the horizontal sweep of the scope, obtain a stable waveform on the CRT screen, displaying a few complete cycles. Sketch the waveform in your lab notebook. Reverse the diode and repeat.

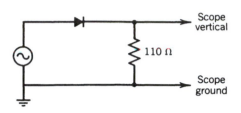

Figure 5

Question 6

Briefly explain the shape of each waveform. You may want to refer to your measurements in parts A and B.

To display the I-V characteristic curve of the diode connect the circuit shown in Figure 6. As before, the horizontal axis of the CRT screen displays the voltage across the device and the vertical axis of the screen is proportional to the current through the device. Vary the oscillator amplitude control knob and the vertical gain control of the scope until the vertical and horizontal displacements of the beam spot are about equal. Observe the curve and sketch it in your notebook. Reverse the diode and repeat.

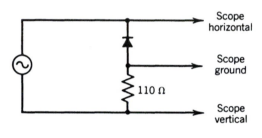

Figure 6

Question 7

Is the I-V characteristic curve observed on the scope in general agreement with the curve from part B? (As before, do not ignore the relative phase shift of 180° due to the chosen ground point.)

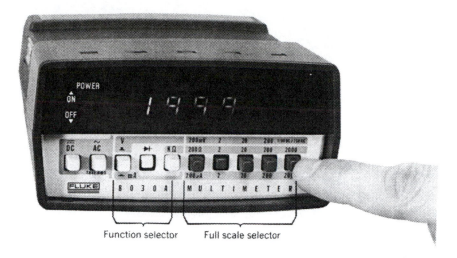

Figure 7

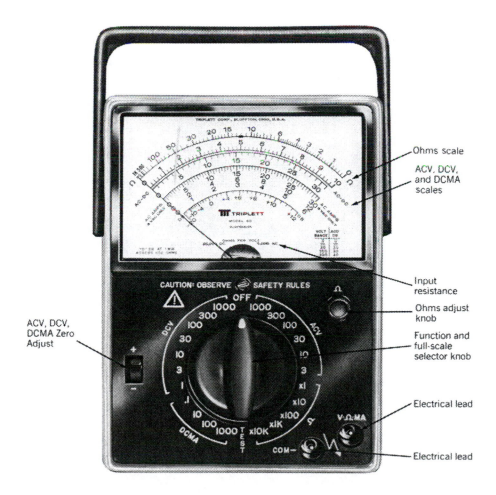

Figure 8

15

AC BEHAVIOR OF RESISTORS, CAPACITORS, AND INDUCTORS

Apparatus

Oscilloscope, oscillator, multimeter, decade resistance box, 10-millihenry inductor, 0.1-microfarad capacitor, 510-Ω ½-watt resistor, 50,000-Ω ½-watt resistor, electrical leads.

Introduction

In part C of Experiment 14, you found the voltage $V_R(t)$ across a resistor is related to the current $I(t)$ by

$$V_R(t) = RI(t) \tag{1}$$

where $V_R(t)$ and $I(t)$ are in phase. We may describe a sinusoidal current by

$$I(t) = I_0 \sin 2\pi\nu t \tag{2}$$

and $V_R(t)$ is then

$$V_R(t) = RI_0 \sin 2\pi\nu t \tag{3}$$

For a pure resistor (a resistor with no capacitance or inductance) R is constant, independent of the frequency ν. Equations (2) and (3) are sketched in Figures 1a and 1b.

The voltage $V_C(t)$ across the plates of a capacitor is related to the charge $q(t)$ on each plate ($+q(t)$ on one plate and $-q(t)$ on the other) and the capacitance C by

$$V_C(t) = \frac{q(t)}{C} \qquad (4)$$

The charging current is related to the charge by $I = dq/dt$ and using Equation (2) we may calculate $q(t)$:

$$q(t) = \int_0^t I(t)\ dt = \int_0^t I_0 \sin 2\pi\nu t\ dt$$

$$= \frac{-I_0}{2\pi\nu} \cos 2\pi\nu t \qquad (5)$$

Thus $V_C(t)$ may be written as

$$V_C(t) = -\frac{I_0}{2\pi\nu C} \cos 2\pi\nu t \qquad (6)$$

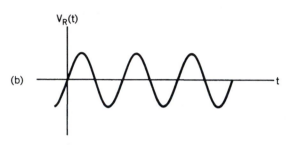

Equation (6) is sketched in Figure 1c. Note that $I(t)$ reaches its maximum one quarter of a period earlier than $V_C(t)$; we say $I(t)$ leads $V_C(t)$ by one quarter of a period or 90°. Also the phase difference between $I(t)$ and $V_C(t)$ may be readily shown by using a well known trig identity

$$\sin(\theta \pm \phi) = \sin\phi \cos\theta \pm \cos\phi \sin\theta \qquad (7)$$

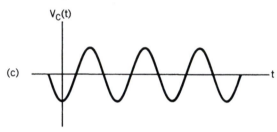

Then Equation (6) may be written

$$V_C(t) = \frac{I_0}{2\pi\nu C} \sin\left(2\pi\nu t - \frac{\pi}{2}\right) \qquad (8)$$

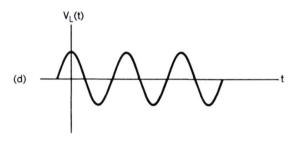

Comparison of Equations (2) and (8) indicates that $I(t)$ leads $V_C(t)$ by a phase angle of $\pi/2$.

Figure 1

The rms voltage across a capacitor may be calculated using Equation (6):

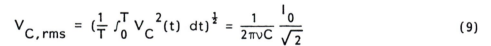

$$V_{C,rms} = \left(\frac{1}{T}\int_0^T V_C^2(t)\ dt\right)^{\frac{1}{2}} = \frac{1}{2\pi\nu C}\frac{I_0}{\sqrt{2}} \qquad (9)$$

The rms current may be calculated using Equation (2):

$$I_{rms} = \left(\frac{1}{T}\int_0^T I^2(t)\ dt\right)^{\frac{1}{2}} = \frac{I_0}{\sqrt{2}} \qquad (10)$$

Substituting Equation (10) into (9), we have

$$V_{C,rms} = \frac{1}{2\pi\nu C} I_{rms} \equiv X_C I_{rms} \tag{11}$$

where X_C has dimensions of resistance and is frequency dependent. It is called the capacitive reactance (AC analog to resistance).

The voltage $V_L(t)$ across the coils of an inductor is given by

$$V_L(t) = L \frac{dI}{dt} \tag{12}$$

where L is the inductance and I(t) is the current through the inductor. Using the sinusoidal current given by Equation (2), V_L becomes

$$V_L(t) = L\, 2\pi\nu I_0 \cos 2\pi\nu t \tag{13}$$

Equation (13) is sketched in Figure 1d. Note that I(t) reached its maximum one-quarter of a period later than $V_L(t)$; we say I(t) lags $V_L(t)$ by one-quarter of a period or 90°. As before, we may use a trig identity, Equation (7), to write $V_L(t)$ in terms of the sine function. The result is

$$V_L(t) = L\, 2\pi\nu I_0 \sin\left(2\pi\nu t + \frac{\pi}{2}\right) \tag{14}$$

Comparing Equations (2) and (14), we find I(t) lags $V_L(t)$ by a phase angle of $\pi/2$.

The rms voltage may be calculated using Equation (13):

$$V_{L,rms} = \left(\frac{1}{T} \int_0^T V_L^2(t)\, dt\right)^{\frac{1}{2}} = 2\pi\nu L\, \frac{I_0}{\sqrt{2}} = 2\pi\nu L\, I_{rms} \equiv X_L I_{rms} \tag{15}$$

where X_L is called the inductive reactance (inductor's AC analog to resistance). X_L has dimensions of resistance and it is frequency dependent.

Table 1 is a summary of the above.

Table 1

Device	V(t)	V_{rms}	Phase
Resistor	$V_R(t) = R\, I(t)$	$V_{R,rms} = R\, I_{rms}$	I(t) & $V_R(t)$ in phase
Capacitor	$V_C(t) = q(t)/C$	$V_{C,rms} = X_C I_{rms}$	I(t) leads $V_C(t)$ by 90°
Inductor	$V_L(t) = L\, dI/dt$	$V_{L,rms} = X_L I_{rms}$	I(t) lags $V_L(t)$ by 90°

In the first part of this experiment you will study the dependence of R, X_C, and X_L on the frequency ν. In the second part you will study the phase of the voltage across one device relative to that of another.

Outcomes

After you have finished the activities in this experiment you will have:

a. studied the frequency dependence of the inductive reactance X_L and the capacitive reactance X_C.

b. studied the phase difference between current and voltage for both a capacitor and an inductor, and observed the phase difference of the voltages across a capacitor and an inductor.

Frequency Dependence of R, X_C, and X_L

The circuit to be used in parts A, B, and C is shown in Figure 2, where the component is a resistor, capacitor, or inductor. We will choose the component such that its resistance or reactance is small compared to the 50 kΩ of the series resistor, hence the rms current may be assumed constant.

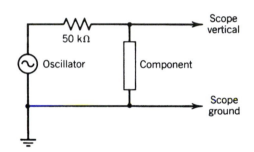

Figure 2

A. Resistance R

Part A is qualitative. Observe the waveform and answer the questions. Recording data is not required. Connect the circuit in Figure 2 with the variable resistance box as the component, set to 1000 Ω. Set the oscillator frequency to 1000 Hz, the amplitude to maximum (leave it set) and obtain a stable waveform on the scope. Vary the frequency of the oscillator from 1000 Hz to 10,000 Hz and observe the voltage across the resistor.

Question 1

Does the resistance depend on frequency? (Recall that I_{rms} or I_0 is assumed constant.)

Set the frequency of the oscillator at 5000 Hz. Vary the resistance from 100 Ω to 1000 Ω, while observing the voltage across the resistance box.

Question 2

Does Equation (1) appear qualitatively to be satisfied?

B. Capacitive Reactance X_C

Replace the resistance box with a 0.1-microfarad capacitor. First you will determine I_{rms}. Set the oscillator frequency at 4000 Hz. Use the scope to measure one or more peak-to-peak voltages which will allow you to calculate the rms current through the 50-kΩ resistor. Do not alter the scope ground connection.

With the circuit connected as shown in Figure 2, vary the oscillator frequency from 1000 Hz to 8000 Hz in 1000-Hz increments and measure V_{pp} across the capacitor. For each frequency calculate the capacitive reactance X_C, using Equation (11). X_C is defined to be $1/2\pi C\nu$. Plot X_C versus some function of the frequency ν such that a straight line results.

Question 3

From the slope of your line, what is the value of C? What is the percent discrepancy between this value of C and that specified by the manufacturer, 0.1 microfarads?

C. Inductive Reactance X_L

Replace the capacitor by a 10-millihenry inductor. Vary the frequency from 1000 Hz to 8000 Hz in 1000-Hz increments and measure V_{pp} across the inductor. Knowing I_{rms} from part B, calculate the inductive reactance X_L for each frequency using Equation (15). X_L is defined to be $2\pi L\nu$. Plot X_L versus ν.

Question 4

What is the value of L determined from the slope of your line? Calculate the percent discrepancy between your experimentally determined value of L and the manufacturer's specified value.

Phase Observation

In parts D, E, and F you will observe Lissajous patterns in order to determine the phase difference between two voltages. It is recommended that you review part E of Experiment 13 before coming to lab.

D. RL Circuit

An RL circuit contains a resistor and an inductor in series. Connect the circuit shown in Figure 3. Note that the oscillator is "floating," i.e., it is not connected directly to ground. Obtaining equal horizontal and vertical displacements of the electron beam spot on the CRT screen is somewhat simplified if the reactance X_L is 510 Ω, equal to the resistance of the resistor. Calculate the frequency required for X_L = 510 Ω. Set the oscillator to that frequency. Vary the oscillator amplitude and the vertical input gain control knob on the scope until approximately equal horizontal and vertical displacements of the beam spot are obtained. Sketch the Lissajous pattern in your notebook.

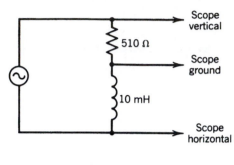

Figure 3

Question 5

What is the magnitude of the phase difference between I(t) and $V_L(t)$? (Refer to Figure 7, Experiment 13 if necessary.) From the observed Lissajous pattern can you determine if I(t) leads or lags $V_L(t)$?

Disconnect the scope from the circuit shown in Figure 3. Using the multimeter as an AC voltmeter, connect it across the oscillator terminals and adjust the oscillator output to 3 V-rms. Then use the multimeter to measure the voltage across the resistor, followed by a measurement of the voltage across the inductor.

Question 6

Why is the voltage across the resistor plus the voltage across the inductor not equal to 3 V-rms?

E. RC Circuit

Replace the 10-millihenry inductor in Figure 3 with a 0.1-microfarad capacitor. Repeat the observations and measurements. Answer questions 5 and 6 for the RC circuit.

F. RLC Circuit

Connect the RLC circuit shown in Figure 4. Set the frequency of the oscillator at 5000 Hz, vary the oscillator amplitude and the vertical gain control of the scope until the vertical and horizontal beam spot displacements are about equal. Sketch the Lissajous pattern in your lab notebook.

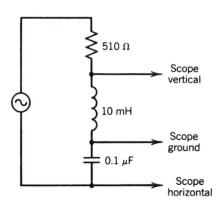

Figure 4

Question 7

From Table 1, what do you expect the phase difference to be? From the Lissajous pattern, what is the observed phase difference? Can you explain this apparent inconsistency between expected and observed phase difference. Hint: With the common point of the inductor and capacitor connected to ground, if V_L is positive (relative to ground), then V_C is negative (relative to ground); i.e., the ground connection causes a 180° phase shift.

Disconnect the scope horizontal from the circuit, switch the horizontal sweep to INTERNAL, and set the SEC/DIV knob to 1 ms. With the frequency still set to 5000 Hz, set the amplitude of the oscillator so that $2 V_{pp}$ is observed across the inductor. Next, connect the scope across the capacitor and inductor, so that it displays the voltage across the series combination and then vary the frequency until V_{pp} across the series combination is a minimum. Leaving the oscillator set, use the scope to first measure $V_{L,pp}$ and then measure $V_{C,pp}$.

Question 8

What do you conclude about the relative phases of the voltages across the inductor and the capacitor?

The frequency which produces the minimum peak-to-peak voltage across the LC series combination is called the <u>resonance frequency</u>. Knowing the relationship between $V_C(t)$ and $V_L(t)$ at "resonance," then the resonance frequency may be calculated.

Question 9

What is the percent discrepancy between your calculated resonance frequency and the value read from the dial of the oscillator?

16

RLC CIRCUITS. DAMPED OSCILLATIONS, DRIVEN OSCILLATIONS, FILTERING

Apparatus

Sine wave oscillator, square wave oscillator, multimeter, oscilloscope, 10-mH inductor, 22-μH inductor, 0.01-μF capacitor, 0.22-μF capacitor, decade resistance box, electrical leads.

Introduction

In this experiment you will study three RLC circuits. The first circuit will be used to study the damped oscillations of electric charge. These electric charge oscillations are analogous to the damped harmonic oscillations of a mass on a spring which you studied in Experiment 8. The second circuit is an RLC series combination with a sinusoidal source of emf. Using this circuit you will study driven oscillations and resonance analogous to the driven oscillations and resonance of the mechanical oscillator studied in Experiment 9. This circuit passes a particular desired frequency and is similar to the "tuned circuit" portion of a radio receiver. Finally you will study an RLC circuit which filters out an unwanted frequency.

An important part of this experiment is the analogy between electrical and mechanical systems from both mathematical and physical points of view. Previous mechanical experiments are referenced in this experiment and it is recommended that you review them before coming to lab.

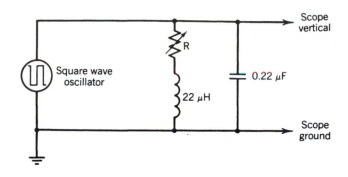

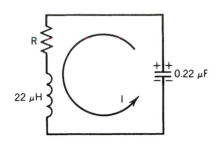

Figure 1   Figure 2

Damped Oscillations of Electric Charge

The square wave oscillator provides a convenient way to repetitively start charge oscillations in the circuit shown in Figure 1. When the oscillator switches from "on" to "off" the capacitor begins to discharge through the RL combination as shown in Figure 2 and the charge will oscillate during the oscillator's "off" time. The resistance R is the resistance of the inductor and the variable resistor. (The internal resistance of the oscillator is large in comparison.) Applying Kirchhoff's loop rule to Figure 2:

$$\frac{q}{C} - L\frac{dI}{dt} - RI = 0 \tag{1}$$

where the positive direction of current is shown in Figure 2, q is the charge on capacitor C, and $I = -dq/dt$. Hence Equation (1) may be written:

$$L\frac{d^2q}{dt^2} + R\frac{dq}{dt} + \frac{q}{C} = 0 \tag{2}$$

The solution to this equation depends on the values of R, L, and C. We will only consider the underdamped case where

$$\frac{1}{LC} > \left(\frac{R}{2L}\right)^2 \tag{3}$$

The solution to Equation (2) is then

$$q(t) = q_0 e^{-Rt/2L} \cos{(\omega't + \phi)} \tag{4}$$

where q_0 and ϕ are constants determined by the initial conditions and ω' is the <u>natural frequency</u>:

$$\omega' = \sqrt{\frac{1}{LC} - \left(\frac{R}{2L}\right)^2} \qquad (5)$$

The period T' is

$$T' = \frac{2\pi}{\omega'} = 2\pi / \sqrt{\frac{1}{LC} - \left(\frac{R}{2L}\right)^2} \qquad (6)$$

and the amplitude is

$$\text{amplitude} = q_0\, e^{-Rt/2L} \qquad (7)$$

The voltage across the capacitor $V_C(t)$, during charge oscillations, is given by

$$V_C(t) = \frac{q(t)}{C} = \frac{q_0}{C}\, e^{-Rt/2L} \cos(\omega't + \phi) \qquad (8)$$

Figure 3 is a photograph of the CRT screen displaying V_C versus t.

Note that if R = 0, then, denoting the angular frequency by ω (analogous to the undamped harmonic oscillator, Experiment 8), we have

$$\omega = \sqrt{\frac{1}{LC}} \qquad (9)$$

and the amplitude is constant, q_0.

You should compare Equations (2) through (7) with their mechanical analogues given in Equations (11) through (15) of Experiment 8. Table 1 is a summary of an electrical system and its mechanical analog.

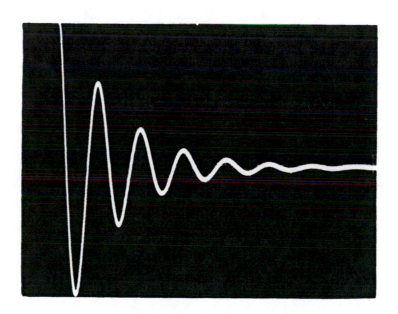

Figure 3

	Table 1		
Electrical System		**Mechanical Analog**	
inductance	L	mass	m
resistance	R	damping constant	b
capacitance	C	(spring constant)$^{-1}$	k^{-1}
charge	q(t)	position	x(t)
current	I(t)	velocity	v(t)
$V_L = L\, dI/dt$		$F = m\, dv/dt$	
$V_C = q/C$		$F_S = kx$	
$V_R = RI$		$F_d = bv$	

Note that each electrical component has a behavior identical to that of its mechanical analogue and vice versa. We can use our knowledge of one system to predict the behavior of the other. This technique of electrical modeling physical systems is widely used in physics and engineering.

Forced Oscillations and Resonance

The RLC circuit in Figure 4 has a sinusoidal driving voltage V_D which we describe by

$$V_D(t) = V_0 \cos \omega''t \qquad (10)$$

where ω'' is the driving frequency and in general it is not equal to the damped frequency ω' or the undamped frequency ω.

Kirchhoff's loop rule applied to the circuit of Figure 4 yields

$$L\frac{d^2q}{dt^2} + R\frac{dq}{dt} + \frac{q}{C} = V_0 \cos \omega''t \qquad (11)$$

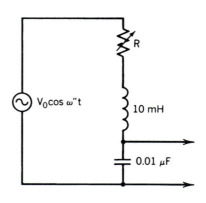

Figure 4

Equation (11) is analogous to Equation (21), Experiment 9, for the forced harmonic oscillator with damping. The solution to Equation (11) is analogous to the solution of Equation (21), Experiment 9.

$$q(t) = q_0(\omega'') \cos (\omega''t + \phi(\omega'')) \qquad (12)$$

where

$$\tan [\phi(\omega'')] = \frac{R\omega''}{L(\omega''^2 - \omega^2)} \qquad (13)$$

$$q_0(\omega'') = \frac{-V_0/L}{\sqrt{(\omega''^2 - \omega^2)^2 + \dfrac{R^2\omega''^2}{L^2}}} \tag{14}$$

You should compare Equations (10) through (14) with Equations (19) through (24), Experiment 9.

We can use Equation (12) and its derivatives to calculate V_C, V_R, and V_L:

$$V_C(t) = \frac{1}{C}\, q(t) = \frac{-V_0/LC}{\sqrt{(\omega''^2 - \omega^2)^2 + \dfrac{R^2\omega''^2}{L^2}}}\ \cos(\omega''t + \phi) \tag{15}$$

$$V_R(t) = R\,\frac{dq}{dt} = \frac{\omega''RV_0/L}{\sqrt{(\omega''^2 - \omega^2)^2 + \dfrac{R^2\omega''^2}{L^2}}}\ \sin(\omega''t + \phi) \tag{16}$$

$$V_L(t) = L\,\frac{d^2q}{dt^2} = \frac{L\omega''^2V_0/L}{\sqrt{(\omega''^2 - \omega^2)^2 + \dfrac{R^2\omega''^2}{L^2}}}\ \cos(\omega''t + \phi) \tag{17}$$

Note the dependence of the amplitudes on ω''. Each amplitude becomes maximum when $\omega'' = \omega$. ω is called the resonance frequency and this maximizing of the amplitudes is called resonance.

Outcomes

After finishing the activities in this experiment, you will have:

a. studied damped oscillations of electric charge.

b. studied driven oscillations and resonance.

c. studied a circuit which filters out an unwanted frequency.

d. seen the correlation between electrical systems and their mechanical analogues.

A. Damped Oscillations

Connect the circuit in Figure 1. R represents the resistance to the left of C, primarily that of the inductor and variable resistor. To measure R, turn off the oscillator and connect the multimeter, used as an ohmmeter, across the capacitor.

Set the oscillator amplitude to about half-maximum, the frequency to 500 Hz (then the square wave period is long compared to T'), and initially set the variable resistor to zero ohms. Obtain a stable waveform on the CRT screen similar to Figure 3. Increase the variable resistor and observe the waveform.

Question 1

What effect does changing R have on the period? What effect does it have on the exponential envelope?

Reset the variable resistor to zero ohms. Measure the period and the time τ required for the amplitude to decay to $1/e$ of its initial value. Calculate the period using Equation (6) and τ, where from Equation (7), $\tau = 2L/R$.

Question 2

What are the percent discrepancies between measured values of the period and τ and their calculated values.

B. Driven Oscillations and Resonance

Connect the circuit in Figure 4. The scope will display the voltage across the capacitor, i.e., it will display Equation (15). Set R to 20 Ω, V_0 to 1 V peak-to-peak and keeping V_0 constant sweep the sine-wave oscillator frequency $\nu''(\omega''/2\pi)$ through the resonance frequency $\nu(\omega/2\pi)$ while observing V_C versus t on the CRT screen. (As you change ν'', V_0 will change and you will have to reset it to 1 V_{pp}.)

Question 3

What is the percent discrepancy between the resonance frequency read from the oscillator and the value calculated from Equation (9). (A source of the discrepancy may be an uncalibrated oscillator.)

Set R to 200 Ω, sweep ν'' through resonance while keeping V_0 constant and observe V_C versus t. Increasing R causes a decrease in the response of the circuit to the driving voltage.

Reset R to 20 Ω. Keeping V_0 constant, measure the peak-to-peak voltage $V_{pp,C}$ across the capacitor as a function of ν''. Plot $V_{pp,C}$ versus ν''. Repeat for R = 200 Ω.

The sharpness of the response to the driving frequency is of interest. The width $\Delta\nu$ of such curves is measured between the two points where $V_{pp,C}$ is $1/\sqrt{2}$ times its maximum value. The quality factor QF is a measure of the sharpness of the response and is given by

$$QF = \frac{\nu}{\Delta\nu} \tag{18}$$

where ν is the resonance frequency.

Question 4

What is the quality factor for each curve? (In microwave resonance circuits quality factors of 10^4 are easily obtained.)

For the circuits considered here $\Delta\nu = R/2\pi L$. Can you show this?

C. Filtering an Unwanted Frequency

Connect the circuit shown in Figure 5. Displaying a stable waveform on the scope, observe V_{pp} across the 1000 Ω as the oscillator frequency is swept through resonance.

Question 5

What is the percent discrepancy between the resonance frequency read from the oscillator dial and the value calculated from $\omega = 1/\sqrt{LC}$?

Question 6

At low frequencies does the capacitor or inductor primarily pass the current? What about at high frequencies?

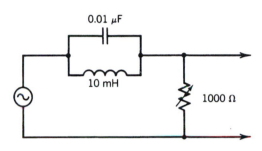

Figure 5

Note the inductor and capacitor are in parallel, hence $V_L(t)$ and $V_C(t)$ are in phase and of equal magnitudes. As before (Experiment 15), the capacitor current $I_C(t)$ leads $V_C(t)$ by 90° and the inductor current $I_L(t)$ lags $V_L(t)$ by 90°.

Question 7

Why is the current through the resistor very small at resonance?

17

INPUT AND OUTPUT RESISTANCES. RESISTANCE MATCHING

Apparatus

Old 1.5-V D cell, new 1.5-V D cell, unregulated DC power supply, oscillator, 1000-Ω $\frac{1}{2}$-watt resistor, decade resistance box, 1 M-ohm $\frac{1}{2}$-watt resistor, VTVM, digital voltmeter, multimeter, oscilloscope, electrometer. (Many of the instruments may be shared by students.)

Introduction

Thevenin's theorem of linear circuit analysis says that at any two points in a circuit the effect of the elements in the circuit is equivalent to an ideal voltage source ε in series with an <u>output impedance</u> Z_0. The two points could be the output terminals of a battery, power supply, oscillator, amplifier, etc., or just two points in some circuit between which we measure the voltage or connect something. Often the resistive part of the output impedance dominates, i.e., $Z_0 \cong R_0$, hence an <u>output resistance</u>. See Figure 1.

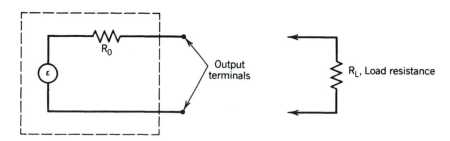

Figure 1

153

If no load resistance is connected, the current is zero (hence zero voltage drop across R_0), and the output voltage (voltage across the output terminals) equals ε. With a load resistance (meter, motor, oscilloscope, etc.) connected the current is not zero and the output voltage is:

$$V = \varepsilon - IR_0 \qquad (1)$$

Also the output voltage is the voltage drop across R_L:

$$V = IR_L \qquad (2)$$

Equating Equations (1) and (2) and solving for I yields:

$$I = \frac{\varepsilon}{R_0 + R_L} \qquad (3)$$

Substituting I from Equation (3) into Equation (1) gives:

$$V = \varepsilon \frac{R_L}{R_0 + R_L} \qquad (4)$$

From Equation (4), note that:

$R_L = 0$ implies $V = 0$ "short circuit current"

$R_L = \infty$ implies $V = \varepsilon$ "open circuit voltage"

The output voltage V versus the output current I is sketched in Figure 2.

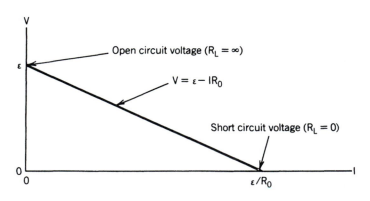

Figure 2

For many circuits and power sources V is a linear function of I only for small output currents. A battery would perhaps burn out and a power supply blow a fuse if the output terminals were short circuited, i.e., R_L = 0. In general power sources are operated in the linear, small-current region and R_0 is the internal or output resistance when operated in this region. In Figure 3 a typical V vs. I curve is sketched.

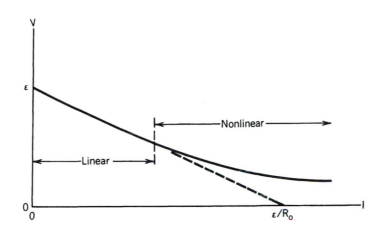

Figure 3

In order to use electrical apparatus properly an understanding of output resistance is necessary. The following two examples illustrate this.

(a) <u>Power transfer.</u> The power delivered to the load resistance R_L is

$$P = I^2 R_L \tag{5}$$

Using Equation (3) to eliminate the current I in Equation (5) yields

$$P = \left(\frac{\varepsilon}{R_0 + R_L}\right)^2 R_L \tag{6}$$

Often it is desirable to obtain maximum power from the source.

Question 1

Using differential calculus, what value of R_L (in terms of R_0) maximizes the power obtained from the source? This is called <u>impedance matching</u> of the source to the load.

Question 2

When maximum power is being transferred from the source to the load, then the output voltage V is equal to what fraction of the open-circuit value ε?

(b) <u>Voltage drop</u>. If a meter is used to measure the voltage drop between two points in a circuit, then you are interested in the value of the voltage with the meter absent. In this case you want $R_L \gg R_0$, where R_L is now the <u>input resistance</u> of the meter. Hence to properly measure the voltage drop you must know the output resistance R_0 of the power source and select a meter such that $R_L \gg R_0$.

Outcomes

After finishing the activities in this experiment you will have:

a. determined the <u>output resistance</u> of an old 1.5-V D cell, a new 1.5-V D cell, an unregulated DC power supply, and an AC oscillator.

b. measured the voltage drop between two points in a circuit using several voltmeters, each having a different <u>input resistance</u>.

c. gained some understanding of output resistance, input resistance, and their role in the proper use of electrical apparatus.

Output Resistance

For each power source connect the circuit shown in Figure 4. The 1000-Ω resistor is to protect the power source, i.e., to prevent a short circuit. In this circuit the load resistance R_L is the 1000 Ω plus the variable resistance R.

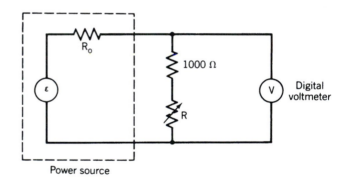

Figure 4

Set R to 9000 Ω and measure V. Decrease R by increments of 1000 Ω until R is 0 Ω, reading V for each value of R. For each value of V and R_L calculate the current I through the load resistance. Plot V vs. I for each power source. For the DC power supply set the voltage to 5 V when $R_L = \infty \ \Omega$ and leave it during your measurements. For the AC power source set the frequency to 5000 Hz and the rms voltage to 5 V when $R_L = \infty \ \Omega$ and leave it.

It is worthwhile to record and plot data in the nonlinear region of the V vs. I curve for the unregulated DC power supply. See your instructor about removing the 1000-Ω resistor and measuring V for values of R less than 1000 Ω.

Question 3

From the slope of each graph determine the output resistance of each power source.

Question 4

If the manufacturer's value of the output resistance is available, what is the percent discrepancy between your value and the manufacturer's value for each power source?

Question 5

If the AC power source is to drive a speaker, what should the input resistance of the speaker be in order that maximum power be transferred to the speaker?

Speakers typically have input resistances of a few ohms, e.g., 4 Ω. An impedance matching transformer is a device which "matches" input and output and hence insures maximum power transfer.

Input Resistance

Connect the new 1.5-V D cell as shown in Figure 5. Measure the voltage V with each voltmeter. Your instructor will provide the manufacturer's specified input resistance for each voltmeter. Voltmeters which may be available are: digital voltmeter, VTVM, multimeter, oscilloscope and an electrometer.

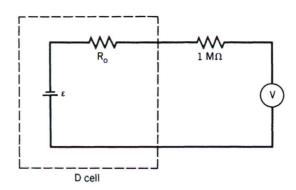

Question 6

Why do the voltmeters give different values of voltage?

Figure 5

18

SEMICONDUCTOR DIODES. DC POWER SUPPLY

Apparatus

Stepdown transformer (110-V rms to 12-V rms), four 1N4001 diodes, one 1N937 zener diode, 1000-ohm $\frac{1}{2}$-watt resistor, 100-ohm $\frac{1}{2}$-watt resistor, 10-μf capacitor, two 100-μf capacitors, oscilloscope, variable resistance box.

Introduction

The vacuum tube was developed in 1907 and for the following 41 years it was used to amplify and rectify AC voltages. In the 1940's a major technological breakthrough occurred: pure semiconductor crystals were grown in the laboratory. This marked the beginning of a sharp increase in research activity in solid state physics: the study of semiconductors. Research in solid state physics led to scientific breakthroughs and to the development of the diode and the transistor in 1948. The diode and transistor have largely replaced the vacuum tube due to their inherent advantages: low power dissipation, long life, low cost, and small size.

Materials can be classified according to their electrical resistivity as conductors, semiconductors, or insulators. Table 1 lists some materials and their resistivity. As shown in Table 1, semiconductors have electrical resistivity values that are intermediate between good conductors and insulators.

Some of the solid state devices which are based on the properties of semiconductor crystals are: diodes, transistors, thermistors, and photocells. The theory of such crystals and devices are beyond the scope of this laboratory; however, it is perhaps worth pointing out that some devices are composed of n-type and p-type semiconductor materials or a combination of these materials. An n-type material is a crystal of silicon (Si) or germanium (Ge) which has been "doped" with an element such as phosphorous (P) or arsenic (As), so that the charge carriers involved in electrical conduction are negative (electrons). A p-type material is a crystal of Si or Ge doped with an element such as boron (B) or aluminum (Al), so that conduction takes place via positive "holes" where an electron may be regarded as missing from the crystal structure.

Table 1	
Material	Resistivity (ohm-m)
Conductors:	
Aluminum	2.8×10^{-8}
Copper	1.7×10^{-8}
Steel	1.8×10^{-7}
Semiconductors:	
Germanium	8.9×10^{2}
Selenium	8.0×10^{4}
Silicon	10 to 10^{4}
Insulators:	
Ceramics	$\sim 10^{11}$
Fused quartz	$\sim 10^{16}$

(The resistivity of conductors usually increases with temperature, while that of semiconductors and insulators decreases.)

A diode is a crystal of Si or Ge which contains both n-type and p-type material. The boundary between the materials is called a p-n junction. See Figure 1. The arrow in the circuit symbol for the diode shows the direction of conventional current. If the voltage is applied in such a direction as to drive the charge carriers toward the p-n junction, then conduction occurs. The current-voltage characteristic of a diode (which you measured in Experiment 14) is sketched in Figure 2.

The I-V curve of a diode shows the sensitivity of a diode to the polarity of the voltage across it. This sensitivity to the polarity is used in many ways. In this experiment you will construct a DC power supply, using diodes in the rectification and regulation stages. A block diagram of a DC power supply is shown in Figure 3a. Figure 3b shows the voltages before and after each major stage.

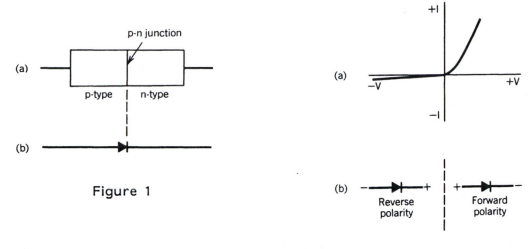

Figure 1

Figure 2

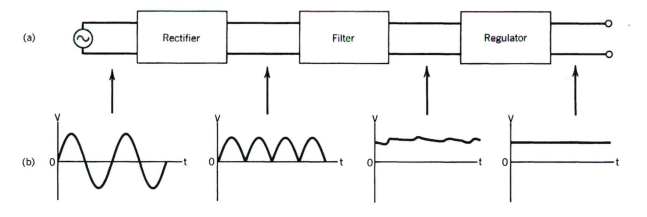

Figure 3

Outcomes

After you have finished the activities you should have a better understanding of:

a. the sensitivity of a diode to the polarity of the voltage across it.

b. the use of diodes to rectify AC voltages.

c. how resistors and capacitors are used to filter AC voltages.

d. the breakdown voltage of a diode and how it is used to regulate the output of a DC power supply (optional).

EXPERIMENTS

A. Half-Wave Rectification

Connect the circuit shown in Figure 4, where the voltage across the secondary of the transformer is 12 V rms. Sketch the waveform observed on the scope, and specify on a circuit diagram the direction of the current through the 1000-Ω resistor. Measure the amplitude of the voltage V_0 and include it in your sketch. Reverse the diode and repeat.

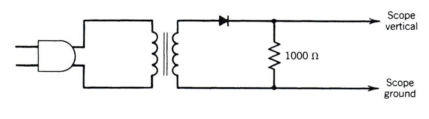

Figure 4

B. Full-Wave Rectification

Connect the "bridge rectifier" circuit shown in Figure 5. Measure the amplitude of the voltage V_0 and sketch the waveform. The polarity of points "a" and "b" is alternating at a frequency of 60 Hz. When "a" is positive then "b" is negative and the conventional current goes from "a" to "b" through the bridge circuit.

Question 1

What is the path of the current through the diodes and 1000-Ω resistor when "a" is positive and "b" is negative? To answer this question draw the circuit and show the current path with arrows. You may assume the scope has infinite input resistance. Repeat for "a" negative and "b" positive.

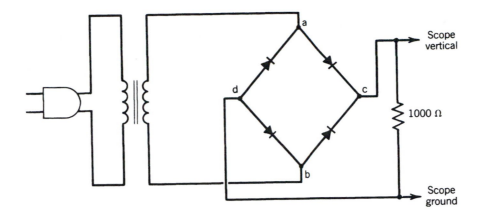

Figure 5

C. Filtering

Connect the circuit shown in Figure 6. The load resistor R_L and the capacitor C form an RC filter. Before connecting the scope to the circuit, set the scope input to DC and short the leads to establish 0 VDC. Then connect the scope across R_L. For C = 10 μf sketch the observed voltage versus time. Your sketch should include DC and AC voltages. See Figure 3b, after the filter stage. Repeat for C = 100 μf.

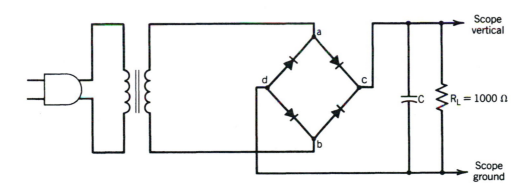

Figure 6

Question 2

Which capacitor produces the smaller "ripple," i.e., the smaller amplitude of the AC voltage across R_L?

For each capacitor calculate the RC time constant for the capacitor and load resistor combination.

Question 3

How does the RC time constant determine the size of the ripple?

Connect the circuit shown in Figure 7. The 100-Ω resistor and the two 100-μf capacitors form an RC π-section filter. Measure the ripple and the DC voltage across the load resistor. The ripple and DC voltage should be smaller than the values measured for the circuit in Figure 6. Thus the RC π-section filter reduces the ripple (that's good) but it also reduces the voltage across the load.

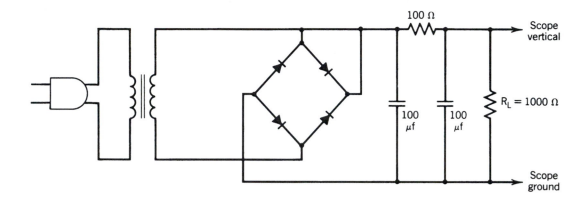

Figure 7

Question 4

Why is the voltage drop across the 1000-Ω load resistance smaller for the circuit in Figure 7 than in Figure 6?

D. Regulation (Optional)

The circuit diagram in Figure 7 is that of an unregulated DC power supply, i.e., the output voltage (voltage across R_L) changes as R_L changes. Verify this by replacing the 1000-Ω load resistor with the variable resistance box PRESET TO 500 Ω. Vary the load resistance from 500 Ω to 3500 Ω in 1000-Ω increments and measure the output voltage for each resistance.

The I-V characteristic curve of a diode drawn in Figure 2a is not complete. If the reverse polarity voltage reaches a critical value, called the "breakdown voltage," then the diode resistance suddenly decreases. See Figure 8. Zener diodes (named after its developer, Mr. Zener) can be manufactured with breakdown voltages from a fraction of a volt to hundreds of volts and with stabilities better than 0.01 percent. Hence they provide excellent reference voltages.

With the variable load resistance PRESET to 500 Ω, connect the circuit shown in Figure 9, which is a regulated DC power supply. The regulated output voltage is determined by the breakdown voltage of the Zener diode. The breakdown voltage of the IN937 Zener diode is nominally 9 V. Vary the load resistance from 500 Ω to 3500 Ω in 1000-Ω increments and measure the DC and AC ripple voltages for each resistance.

Look closely at your measured DC voltages for the regulated and unregulated power supply. From your data the meaning of regulated and unregulated should be clear.

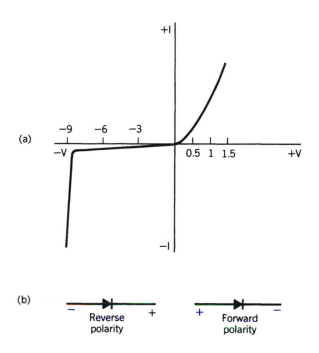

Figure 8

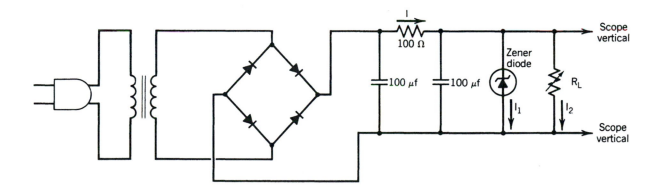

Figure 9

19
BLACK BOXES

Apparatus

Oscillator, oscilloscope, multimeter, electrical leads, decade resistance box, 10-millihenry inductor, 0.1-microfarad capacitor, diode, 50,000-ohm ½-watt resistor.

Introduction

Each black box contains various circuit elements of types you have already studied: resistor, capacitor, inductor, diode and/or a battery. Using techniques from previous labs, you should be able to determine what is in any of the black boxes, except that a small resistance in series with a diode may be impossible to spot (it is there to protect the diode). Possible black boxes are shown in Figure 1.

Outcomes

After you have finished the activities in this exercise you will have:

a. tested your approach to new problems.
b. applied your knowledge of electrical circuits.

Experiment

First check for voltage; there may be a battery inside. If there is, then determine internal resistance using resistors and voltmeter, not ohmmeter. If there is no voltage present, then you can use the ohmmeter to check resistance. For capacitors and inductors you will need to set up an oscillator and scope. It is not necessary to exceed 10^4 Hz. Do not exceed 10 volts peak-to-peak.

An inductor, capacitor and diode are provided as part of the apparatus so that you may review their properties. Make it clear in your notebook (1) why you decided to make that particular measurement, (2) what the measurements were and (3) what conclusion you reached after each measurement. Finally, draw the circuit diagram of the hidden circuit inside the box and show it to the instructor before proceeding to another box. Do as many boxes as you can comfortably. (This is not a contest.)

POSSIBLE BLACK BOXES

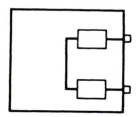

Single component black boxes; resistor, capacitor, or inductor.

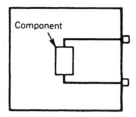

Two components in series: a battery and a resistor, a resistor and a capacitor, a resistor and a diode, or a capacitor and an inductor.

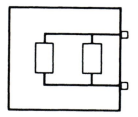

Two components in parallel: a resistor and a capacitor, or an inductor and a capacitor.

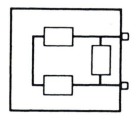

Three components: a diode and a resistor in parallel with a capacitor, or a battery and a resistor in parallel with a capacitor.

Figure 1

20
TRANSISTOR. AC AMPLIFIER

Apparatus

Power supply, transistor #2N1682, 1000-ohm $\frac{1}{2}$-watt resistor, 10000-ohm $\frac{1}{2}$-watt resistor, 2.2-M ohm $\frac{1}{2}$-watt resistor, 0.1-microfarad capacitor, oscillator, VTVM, oscilloscope, electrical leads.

Introduction

The vacuum tube was used to amplify AC voltages from 1907 to 1948. In 1948 the transistor was developed and it essentially replaced the vacuum tube due to its inherent advantages: low power dissipation, long life, low cost, and small size. The integrated-circuit (IC) chip was developed in 1959 and it has essentially replaced the single transistor due to its greater range of capabilities.

The IC chip is a single semiconductor crystal, usually silicon, having suitably disposed n-type and p-type regions. See Experiment 18, SEMICONDUCTOR DIODES. The basic component of IC chips used in radios, televisions, watches, calculators, and computers is the transistor; however, an IC chip does include other components such as diodes and resistors. Since transistors are basic to an IC chip it is worthwhile to examine a circuit using a single transistor.

If they are available in the lab have a look at a vacuum tube (equivalent to a transistor) and an IC chip which contains thousands of transistors. Without the transistor spaceships *Columbia* and *Challenger* would at most not exist and at least be much less sophisticated.

A "junction" transistor is a single semiconductor crystal, usually silicon or germanium, that has two junctions between n-type and p-type regions. An electrical lead connects to each region and the regions are labeled: the emitter (E), the base (B), and the collector (C). There are two types of junction resistors:

npn transistor: p-type region is sandwiched between n-type regions.

pnp transistor: n-type region is sandwiched between p-type regions.

The two types of junction transistors are shown schematically and symbolically in Figure 1.

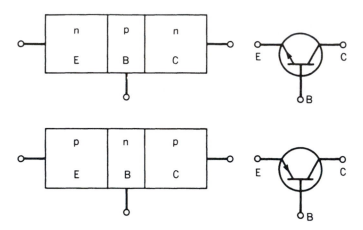

Figure 1

The 2N1682 transistor is an npn transistor. In part A of this experiment the transistor will be connected in a DC circuit which provides suitable operating voltages for the transistor. In part B a small AC signal which is to be amplified is applied to the circuit constructed in A. In part C the gain and bandwidth of the amplifier are investigated and in part D you will examine the relative phase of the input and output signals.

Outcomes

When you have finished the activities in this exercise you will have:

a.   assembled an AC amplifier which uses a transistor.
b.   investigated the amplifier gain and bandwidth.
c.   measured the phase of the output signal relative to the input signal.

A. DC Operating Voltage

Before a signal is applied and amplified, the transistor must be arranged to operate at a suitable set of voltages and currents. Connect the circuit shown in Figure 2. Connect the VTVM probe to CC and adjust the power supply to 15 V. Note at point CC, the total current I_E (emitter current) splits into I_B (base current) and I_C (collector current):

$$I_E = I_C + I_B \tag{1}$$

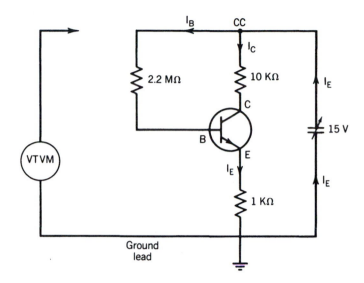

Figure 2

Use the VTVM to measure voltages at E, B, C, and CC. Measure all four voltages with respect to ground. V_C should be between 5 and 10 volts; if not, your transistor or your circuit may be at fault. (A minor adjustment in the 2.2-M resistor may be required to compensate for intrinsic differences in transistors.) Calculate I_B, I_C and I_E. For example,

$$I_C = (V_{CC} - V_C)/10^4 \text{ ohms} \tag{2}$$

You should find that very little current flows through the base, so that the emitter and collector currents are almost equal.

The emitter-base, n-p junction acts essentially as an n-p junction diode. If the emitter voltage V_E is negative relative to the base voltage V_B, then the junction is forward-biased and the transistor passes current (like a forward-biased diode). The larger the forward-bias voltage the greater the current (again like a diode). If the emitter-base junction is reverse-biased ($V_E \geq V_B$), then the transistor does not pass current.

The emitter-base junction in Figure 2 is forward-biased, allowing current to pass through the transistor.

Question 1

From your measured values of V_E and V_B, what is the forward-bias voltage between emitter and base?

To obtain a better understanding of the importance of emitter-base bias voltage, let's now change it and observe its effect on transistor current. Connect the VTVM probe to C. Connect a wire from B to Ground. Observe the effect on the voltage at C. Leaving B connected to Ground connect the VTVM probe to E and observe the voltage.

Question 2

(a) With B connected to Ground what is the forward-bias voltage between emitter and base?

(b) What are the collector and emitter currents when B is connected to Ground?

Remove the wire connecting B to Ground.

Question 3

If the forward-bias voltage of the emitter-base junction is increased, does the collector voltage V_C increase or decrease? What happens to V_C if the forward-bias voltage is decreased?

B. Transistor as an Amplifier

Leave the circuit in part A connected. Connect the oscillator to the
oscilloscope via the 0.1-microfarad capacitor as shown in Figure 3. Adjust the
oscillator amplitude and frequency to
give a sine-wave of 0.5 volts peak-to-
peak, at 1000 Hz. Leave the oscillator
settings and connect the oscillator and
scope to the transistor circuit. See
Figure 4.

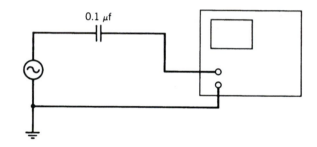

The small AC voltage applied
to the base causes the base voltage to
periodically change. Hence the emitter-
base bias voltage will also change
periodically. See Figure 5.

Figure 3

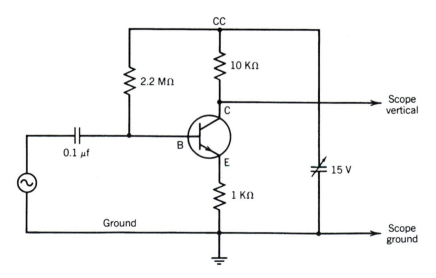

Figure 4

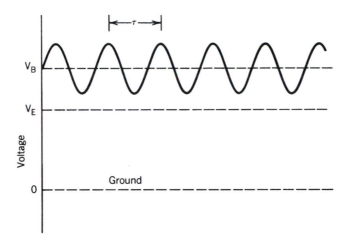

Figure 5

Question 4

Knowing the forward-bias DC voltage across the emitter-base junction which you established in part A and knowing the peak-to-peak amplitude of the AC signal applied to the base, between what voltage range will the forward-bias voltage now swing?

Question 5

How does the periodically changing emitter-base forward-bias voltage effect the collector current and hence the collector voltage?

Observe the waveform of the collector voltage on the scope. Your observation should agree with your answer to Question 5!

Now increase the input signal to 2 volts peak-to-peak.

Question 6

What happens to the output, and why?

C. Amplifier Gain and Bandwidth

The voltage gain of an amplifier is defined as:

$$\text{voltage gain} = \frac{\text{peak-to-peak output (collector) voltage}}{\text{peak-to-peak input (base) voltage}} \qquad (3)$$

Use the scope to verify the peak-to-peak input voltage at the base B is 0.5 V at 1000 Hz. Then alternately measure the peak-to-peak input and output voltages at frequencies of 50, 100, 500, 1000, 5000, 10000, etc. up to 500000 Hz. At each frequency calculate the voltage gain.

On 5-cycle semilog paper plot voltage gain vs. frequency, plotting frequency along the logarithmic axis. Draw a smooth curve through the data points.

Question 7

From the smooth curve, what is the maximum voltage gain?

Question 8

The bandwidth of an amplifier is the frequency range between voltage gains that are equal to the maximum voltage gain divided by $\sqrt{2}$. On your graph locate these two voltage gains and hence determine the bandwidth of your amplifier.

D. Phase

Question 9

What is the phase of the AC signal at the collector relative to that of the small AC signal applied to the base?

To answer Question 9 apply the two signals to the scope and observe the Lissajous pattern.

E. Improving the Gain (Optional)

Connect a 1-microfarad, electrolytic capacitor across the 1000-ohm resistor, from E to ground.

Question 10

With the capacitor connected, what is the voltage gain at 8000 Hz, say?

The theoretical voltage gain is approximately:

$$\text{theoretical gain} \cong \frac{\text{collector resistance}}{\text{emitter resistance}} \qquad (4)$$

where the collector resistance is the resistance which connects to the collector (10 K Ω in this case) and the emitter resistance is the resistance which connects to the emitter (1 K Ω without the capacitor).

Question 11

What is the theoretical gain with and without the 1-microfarad capacitor?

Question 12

What is the percent discrepancy between observed gain at 8000 Hz and the theoretical gain? Do the calculation for the 1-microfarad capacitor connected and removed.

Note that this capacitor improves gain at the expense of bandwidth.

F. Driving a Speaker (Optional)

Suppose the amplifier is to be used to drive a 4-ohm speaker. See Figure 6.

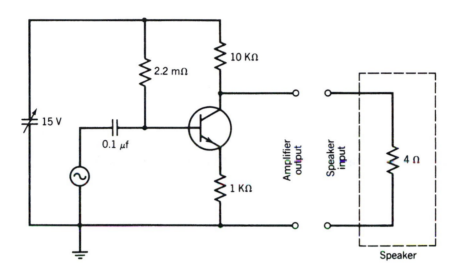

Figure 6

Question 13

If the speaker input was connected directly to the amplifier output, would the DC operating voltages of the amplifier be altered?

Question 14

What circuit component should be used to connect the amplifier to the speaker such that the amplifier's DC voltages are not changed?

21
INTEGRATED CIRCUIT OPERATIONAL AMPLIFIERS

Apparatus

Integrated circuit operational amplifier (Fairchild μA741), +12 V and -12 V DC power supply, oscilloscope, oscillator (sine, square, and sawtooth waveforms), resistors: 270-Ω ¼-watt, 1000-Ω ¼-watt, 9100-Ω ¼-watt, three 10000-Ω ¼-watt, 50000-Ω ¼-watt, 100000-Ω ¼-watt; 0.1-μF capacitor, two 1.5-V D cells, experimentor sockets, electrical leads.

Introduction

A. INTEGRATED CIRCUITS

In previous experiments you studied discrete circuit components: resistors, capacitors, inductors, diodes, and transistors. These components were separately manufactured and you formed electrical circuits by wiring the components together with metallic conducting wire.

An integrated circuit or IC consists of a single crystal chip of silicon containing many circuit components and their necessary connections. So far inductors have not been conveniently fabricated in IC chips, but resistors, capacitors, diodes, and transistors have been. In 1984 IBM produced an experimental computer memory silicon chip that can store more than 1 million bits of information. The storage of each bit requires one or more transistors. The 1-million bit silicon chip developed by IBM measures 3/8 inch by 5/16 inch.

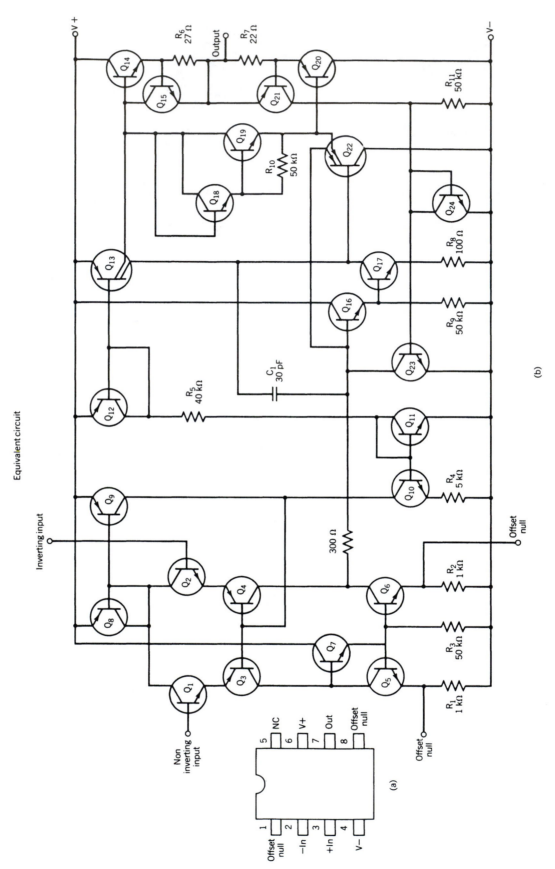

Figure 1

B. INTEGRATED CIRCUIT OPERATIONAL AMPLIFIER

In this experiment you will study several circuits using an IC operational amplifier (op-amp). A top view of the package enclosing the Fairchild μA741 IC op-amp chip is shown in Figure 1a. Figure 1b shows the equivalent circuit diagram. Note that the chip contains: 24 transistors, 11 resistors, and 1 capacitor. The IC op-amp chip is a small-scale IC chip.

The circuit symbol for the op-amp is shown in Figure 2. Depending on the configuration one of the inputs usually connects to ground. The <u>inverting input</u> signifies that the output signal will be 180° out of phase with the input signal applied at this terminal.

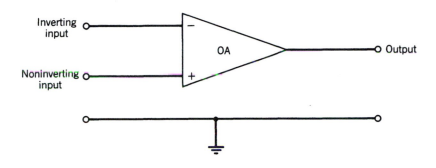

Figure 2

A particular configuration of the op-amp is shown in Figure 3; it is called the inverting configuration because the output signal is inverted (180° out of phase) relative to the input signal. In this experiment you will study several configurations of the op-amp.

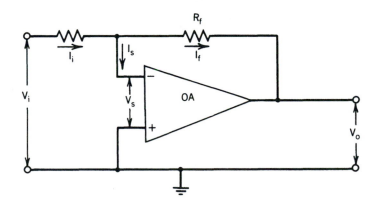

Figure 3

We now consider <u>feedback</u> and <u>gain</u> of a general voltage amplifier. Then we will consider the feedback and gain of an op-amp in the inverting configuration. We start with a voltage amplifier, Figure 4a, that has a voltage gain $A' = V_0'/V_s$. Figure 4b shows the voltage amplifier with feedback: a fraction β of the output voltage V_0 is fed back to the input. How does feedback effect amplifier gain? Well, the output V_0 with feedback is related to V_s by the gain A' without feedback:

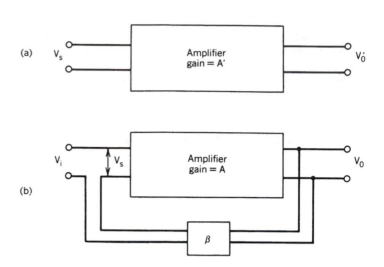

$$V_0 = V_s A' \qquad (1)$$

The signals V_i, V_s, and V_0 are related by β:

$$V_s = V_i \pm \beta V_0 \qquad (2)$$

Figure 4

If the feedback signal βV_0 is in phase with V_i, then β is (+); this is called <u>positive feedback</u>. If the feedback signal βV_0 is 180° out of phase with V_i, then β is (-) and this is called <u>negative feedback</u>. The gain A with feedback is

$$A \equiv \frac{V_0}{V_i} = \frac{V_s A'}{V_s \mp \beta V_0} = \frac{V_s A'}{V_s \mp \beta V_s A'} = \frac{A'}{1 \mp \beta A'} \qquad (3)$$

For positive feedback A approaches infinity as $\beta A'$ approaches one and for negative feedback A is always less than A'. Both types of feedback have their uses.

In Figure 3 the signal is fed back to the inverting input of the op-amp via the "feedback" resistor R_f. The feedback is negative since V_0 and V_s are 180° out of phase.

The op-amp is a high-gain amplifier with a frequency response from DC to about 5 MHz, that is, it will amplify signals having frequencies in this range. The ideal op-amp has very high input impedance, very low output impedance, and very high gain. Typical values are: input impedance ~6M Ω, output impedance ~75 Ω, gain ~50,000. These precise values are usually of little importance in the operation of the complete circuit. The output V_0 is determined by the feedback resistor R_f. This will be shown below.

Applying Kirchhoff's current rule to the circuit in Figure 3 we have

$$I_i = I_s + I_f \qquad (4)$$

For the ideal op-amp the input impedance is very high; therefore, I_s is very small. We assume $I_s \cong 0$ and then

$$I_i \cong I_f \qquad (5)$$

The potential difference $V_i - V_s$ across R_i is given by

$$V_i - V_s = R_i I_i \tag{6}$$

and the potential difference $V_s - V_0$ across R_f is

$$V_s - V_0 = R_f I_f \tag{7}$$

For the ideal op-amp the gain is very high; therefore, $V_0 >> V_s$. We assume the potential difference across R_i so large that V_s is negligible compared to V_i. Then we may write Equations (6) and (7) as

$$V_i \cong R_i I_i \tag{8}$$

$$-V_0 \cong R_f I_f \tag{9}$$

Combining Equations (5), (8), and (9), we find the gain A of the op-amp in the inverting configuration is

$$A \equiv \frac{V_0}{V_i} \cong -\frac{R_f}{R_i} \tag{10}$$

The negative sign implies the two signals are 180° out of phase. Note that the gain of the complete circuit does not depend on the particular value of the gain of the op-amp IC but only on the assumption that it has a large gain.

Outcomes

After you have finished the activities in this experiment you will have:

a. been introduced to an IC op-amp chip.

b. been introduced to amplifiers with feedback.

c. studied an IC op-amp used in the following configurations:
    (1) Inverting, (2) Noninverting, (3) Summing, (4) Differentiator,
    (5) Integrator.

A. Inverting Configuration of the Operational Amplifier

The suggested circuit is shown in Figure 3. Connect the circuit with R_i = 1000 Ω, R_f = 10,000 Ω, and the sine wave oscillator as V_i. Set the oscillator at 1000 Hz and use the scope to set $V_{i,pp}$ = 0.1 V. Measure $V_{0,pp}$ with the scope and calculate the op-amp gain A. Replace the 10,000-Ω resistor with a 50,000-Ω resistor and repeat. For each resistor calculate the ideal gain given by Equation (10).

Question 1

For each feedback resistor what is the percent discrepancy between your measured gain and the ideal gain?

With R_f = 50,000 Ω, measure $V_{0,pp}$ at frequencies of 50, 100, 500, 1000, 5000, 10000, 50000, 100000, and 200000 Hz. For each measurement check $V_{i,pp}$ and reset it to 0.1 V as necessary. Plot gain versus frequency on 4-cycle semilog paper, plotting frequency on the logarithmic axis. Determine the bandwidth of the amplifier. Compare the gain and bandwidth of the op-amp with that of the single transistor amplifier in Experiment 20.

Question 2

What is the ratio of the single transistor amplifier gain to that of the op-amp with R_f = 10,000 Ω? Which amplifier has the broader bandwidth?

Design and connect a circuit which will allow you to determine the magnitude of the phase angle between V_0 and V_i. Hint: Observe the Lissajous pattern on the scope. Draw your circuit and the observed Lissajous pattern in your lab notebook.

Question 3

Is V_0 inverted relative to V_i, that is are the two signals 180° out of phase?

Replace the 50,000-Ω resistor with the 10,000-Ω resistor. Disconnect the oscillator from the op-amp input, connect the 1.5-V D cell to the input, and measure the DC output voltage.

Question 4

What is the DC gain of the op-amp?

B. Noninverting Configuration of the Operational Amplifier

The suggested circuit is shown in Figure 5. Note that V_i is applied to the noninverting input of the op-amp. Analysis shows the theoretical gain is

$$A = \frac{R_1 + R_f}{R_1} \qquad (11)$$

Connect the sine wave oscillator to the op-amp input. Set the frequency to 1000 Hz, $V_{i,pp}$ to 0.1 V, then measure $V_{0,pp}$. Calculate the gain.

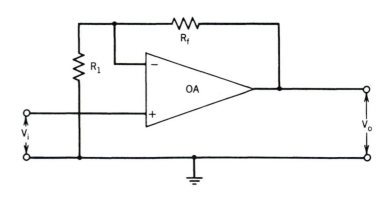

Figure 5

Question 5

What is the percent discrepancy between the gain calculated from your measured values and the theoretical gain given by Equation (11)?

Design and connect a circuit which will allow you to compare the phases of V_0 and V_i. Draw the circuit and the observed Lissajous pattern in your notebook.

Question 6

Are the signals V_i and V_0 noninverted, that is, is the relative phase angle zero?

C. Summing Configuration of an Operational Amplifier

The suggested circuit is shown in Figure 6. Note that there are two input signals, labeled V_1 and V_2. More than two signals could be applied. Analysis shows

$$-\frac{V_0}{R_f} = \frac{V_1}{R_1} + \frac{V_2}{R_2} \qquad (12)$$

Choose $R_1 = R_2 = R_f = 10000 \ \Omega$; then the output voltage is the negative sum of the input voltages, hence, the name summing configuration.

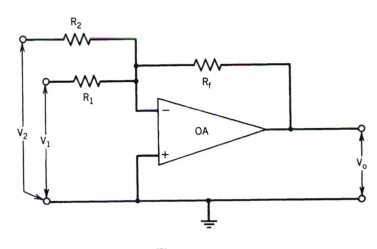

Figure 6

Sum two or more of the following voltages and describe and/or sketch the observed output voltage V_0 in your lab notebook.

(a) V_1 = 1V peak-to-peak, 1000 Hz sine wave

   V_2 = 1V peak-to-peak, 1000 Hz sine wave, then vary the frequency of V_2

(b) V_1 = 1V peak-to-peak, 1000 Hz sine wave

   V_2 = 1V peak-to-peak, 1000 Hz square wave

(c) V_1 = +1.5-V D cell

   V_2 = +1.5-V D cell

(d) V_1 = +1.5-V D cell

   V_2 = -1.5-V D cell

D. Differentiator Configuration of an Operational Amplifier

The suggested circuit is shown in Figure 7. Analysis shows that the output V_0 is proportional to the time derivative of the input dV_i/dt:

$$V_0 = -R_fC\frac{dV_i}{dt} \qquad (13)$$

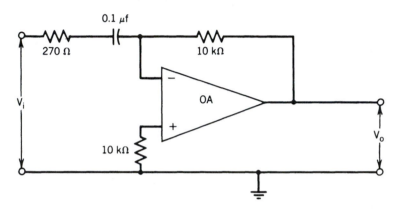

Figure 7

Apply a 2 V_{pp}, 1000 Hz, sawtooth wave to the input and observe the output on the scope. Sketch the input and output waveforms in your notebook, and write cogent remarks pertaining to the agreement of your waveforms and Equation (13).

E. Integrator Configuration of an Operational Amplifier

The suggested circuit is shown in Figure 8. Analysis shows V_0 is proportional to the integral of V_i:

$$V_0 = -\frac{1}{R_fC}\int V_i\,dt \qquad (14)$$

Apply a 4 V_{pp}, 1000 Hz, square wave to the input. Sketch the input and output waveforms in your notebook, and comment on the agreement of Equation (14) with your waveform.

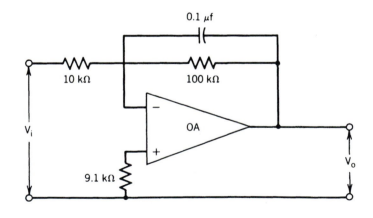

Figure 8

22

REFLECTION AND REFRACTION. DISPERSION

Apparatus

Laser, 2 pins, glass cube, protractor, 90° glass prism, 60° glass prism, spectrometer, mercury arc lamp.

Introduction

Some important predictions of Maxwell's equations involve:

1. propagation of electromagnetic (EM) waves in vacuum and in matter.
2. behavior of EM waves at boundaries.

The electric and magnetic fields of plane EM waves propagating in the x direction in a vacuum are:

$$E_z(x,t) = E_0 \sin (kx - \omega t) \tag{1}$$

$$B_y(x,t) = B_0 \sin (kx - \omega t) \tag{2}$$

where we assumed the electric field lines are in the z direction. E_0 and B_0 are the amplitudes, k is the wavenumber $2\pi/\lambda$, and ω is the angular frequency $2\pi/T$. From Faraday's law the phase velocity v (velocity of the crests and troughs) is given by

$$v = \frac{E_0}{B_0} \tag{3}$$

From the wave equation v is given by

$$v = \frac{1}{\sqrt{\mu_0 \varepsilon_0}} \equiv c \tag{4}$$

where μ_0 is the magnetic permeability of the vacuum and ε_0 is the electric permittivity of the vacuum. In 1974 "c" was reported as 299,792,459.0 ± 0.8 m/s. It is now defined as 299,792,458 m/s (1983).

An ideal dielectric is nonmagnetic and it does not have charge free to flow. Glass and mylar approximate ideal dielectrics. Maxwell's equations which describe a plane EM wave in an ideal dielectric differ from the equations describing the wave in a vacuum by the replacement of ε_0 with ε, the electric permittivity of the dielectric. (The dielectric constant K and the index of refraction n are related to ε and ε_0 by: $K = n^2 = \varepsilon/\varepsilon_0$, where $K > 1$.) Just as the resultant electric field between the plates of a capacitor is reduced when a dielectric is inserted, the amplitude E_0' of an EM wave is reduced in a dielectric. The amplitude B_0 of the magnetic field is not effected by the dielectric. From Faraday's law and the wave equation the phase velocity of the plane EM wave in an ideal dielectric is

$$v = \frac{E_0'}{B_0} = \frac{1}{\sqrt{\mu_0 \varepsilon}} \tag{5}$$

Also $\varepsilon = \varepsilon_0 n^2$, so the phase velocity may be written

$$v = \frac{1}{\sqrt{\mu_0 \varepsilon_0}} \; \frac{1}{n} = \frac{c}{n} \tag{6}$$

and $n > 1$, thus $v < c$.

The angular frequency ω of the wave in the dielectric does not change from the vacuum value; however, the wavelength λ does change. If the wavelength in the vacuum is λ, then in the dielectric it is λ/n.

Summarizing a plane EM wave in a dielectric relative to a wave in the vacuum: the amplitude of the electric field is reduced, the phase velocity is less than c, the frequency does not change, and the wavelength decreases.

The intensity I (energy per unit area per unit time) of an EM wave having amplitude E_0 (or E_0' in a dielectric) and velocity v is

$$I(x,t) = \frac{E_0^2}{v} \sin^2 (kx - \omega t) \tag{7}$$

A detector located at x = 0, say, measures the time-averaged intensity $\bar{I}$:

$$\bar{I} = \frac{1}{T} \int_0^T I(0,t) \, dt = \frac{E_0^2}{2v} \tag{8}$$

We now know about propagation of plane EM waves in a vacuum and in a dielectric. What happens to an EM wave incident on a dielectric from a vacuum? If we shine light on glass, we observe that some light is reflected and some is refracted as shown in Figure 1. We ask

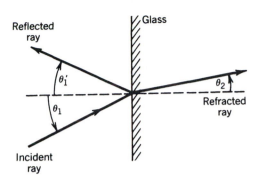

Figure 1

1. How are θ_1' and θ_2 related to θ_1?

2. What fraction of the incident wave energy or intensity is reflected?

3. What are the angular dependencies of the reflected and refracted energies?

The answers come from Maxwell's equations. The incident, reflected, and refracted waves must satisfy Maxwell's equations at the boundary. The results are:

1. Law of reflection: $\theta_1 = \theta_1'$ (9)

2. Snell's law of refraction: $n_1 \sin \theta_1 = n_2 \sin \theta_2$ (10)

3.
$$\frac{\text{Reflected intensity}}{\text{Incident intensity}} = \left(1 - \frac{n_2 \cos \theta_2}{n_1 \cos \theta_1}\right)^2 \Big/ \left(1 + \frac{n_2 \cos \theta_2}{n_1 \cos \theta_1}\right)^2 \tag{11}$$

4.
$$\frac{\text{Refracted intensity}}{\text{Incident intensity}} = 4n_2/n_1 \left(1 + \frac{n_2 \cos \theta_2}{n_1 \cos \theta_1}\right)^2 \tag{12}$$

Recall that intensity is proportional to the wave energy, hence conservation of wave energy requires:

$$\text{Incident intensity} = \text{Reflected intensity} + \text{Refracted intensity} \tag{13}$$

Sum Equations (11) and (12) and then show the sum reduces to Equation (13). In this experiment you will not do intensity measurements.

Figure 2 shows white light incident on glass from air. The glass disperses the light into colors and this phenomenon is called dispersion. The index of refraction of the glass is frequency dependent; n_g increases as the frequency of visible light increases. From Snell's law higher values of n_g mean smaller angles of defraction θ_2 as shown in Figure 2.

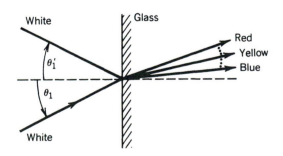

Figure 2

Outcomes

After you have finished the activities in this experiment you will have:

a. verified the laws of reflection and refraction.
b. determined the index of refraction of a prism.
c. determined the critical angle of a prism.
d. determined the dispersive properties of a prism.

A. Laws of Reflection and Refraction

Place the glass cube on a page of your notebook and trace its outline. Arrange a laser so that the rays are similar to those shown in Figure 3. Insert two pins into the page of your notebook to locate each of the three rays which are outside the cube. Trace the three rays on the page and then measure the angles.

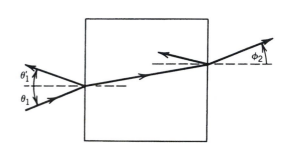

Figure 3

Question 1

Do you find the laws of reflection and refraction are satisfied? Suggestion: Apply Snell's law at each boundary and determine ϕ_2 in terms of θ_1, then compare your measured angles with the prediction.

B. Minimum Deviation Angle

Place the 90° prism on a page of your notebook. Arrange a laser so that the rays are similar to those shown in Figure 4. Rotate the prism until the deviation angle Ψ is a minimum, then trace the rays using pins and trace the outline of the prism. Figure 4 shows the symmetrical geometry of a light ray traversing a prism at minimum deviation. Use Snell's law and geometry to show:

$$n_g \sin \frac{\phi}{2} = n_a \sin \left(\frac{\Psi + \phi}{2}\right) \qquad (14)$$

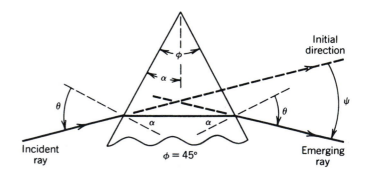

Figure 4

where $\phi = 45°$, and n_g and n_a are the index of refraction of glass and air. Assuming $n_a = 1$, calculate n_g. Look up indices of refraction of glasses in your text or in the Handbook of Chemistry and Physics.

Question 2

Other than experimental errors, why is your value not likely to agree with those in the table?

C. Critical Angle and Total Reflection

The critical angle for the boundary separating two optical media is defined as the smallest angle of incidence, in the medium of greater index of refraction, for which light is totally reflected. The critical angle θ_c and a totally reflected ray are shown in Figure 5. The critical angle is found by putting $\theta_2 = 90°$ into Snell's law:

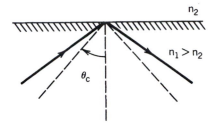

$$n_1 \sin \theta_c = n_2 \sin 90° \qquad (15)$$

or

Figure 5

$$\sin \theta_c = \frac{n_2}{n_1} \qquad (16)$$

Place the 90° prism on a page of your notebook. Direct the laser beam as shown in Figure 6a and observe the reflected ray. θ_1 is greater than the critical angle, hence total internal reflection occurs. Trace the prism and rays in your notebook.

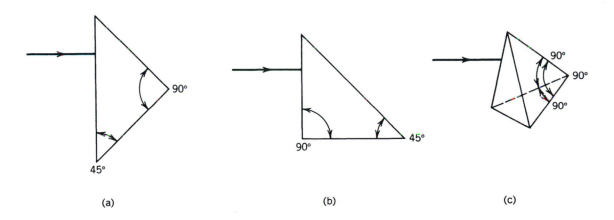

(a)   (b)   (c)

Figure 6

Question 3

Knowing the index of refraction from part B, what is the critical angle for your 90° prism?

Direct the laser beam as shown in Figure 6b. Trace the prism and the path of the ray in your notebook.

Figure 6c is called a "triple mirror" prism. It is made by cutting off the corner of a glass cube by a plane which makes equal angles with the three faces intersecting at that corner.

Question 4

A ray striking any surface of the triple mirror prism will emerge from the prism traveling in what direction? (Hint: the ray is internally reflected from three surfaces. Astronauts left triple mirror prisms on the moon.)

D. Dispersion of a Prism

In this experiment you will use a prism spectrometer, shown in Figures 7 and 8, to determine the index of refraction of a 60° glass prism as a function of wavelength. The prism spectrometer is a device for accurately measuring the angle of deviation of a light ray by a prism. Since the prism disperses the light according to wavelength, the prism spectrometer can be used to determine the spectral composition of the incident light.

The collimator shown in Figure 7 provides parallel light rays which are deviated by the prism through an angle Ψ (see Figures 4 and 7) which depends on the wavelength. The telescope can be rotated so that the angle Ψ of deviation can be measured to the nearest one minute of arc (1/60 degree).

There are three adjustable screws on the spectrometer which you should become familiar with before attempting to do measurements.

1. <u>Slit-width screw</u>. The slit-width screw is located on the side of the collimator near the slit. This screw varies the slit width by moving one side of the slit; hence, it allows you to control the light intensity. Always focus the telescope crosshairs on the nonmovable side of the slit.

2. <u>Telescope tightening screw</u>. The tightening screw is located on the telescope base. When this screw is loose the telescope may be freely rotated. When it is tightened you may change the angular position of the telescope with the fine adjustment screw only.

3. <u>Telescope fine adjustment screw</u>. The fine adjustment screw is also located on the telescope base. With the tightening screw secured the fine adjustment screw may be used to rotate the telescope through small angles. Use the fine adjustment screw to approach a spectral line from the same side; to do otherwise would probably cause an error from backlash.

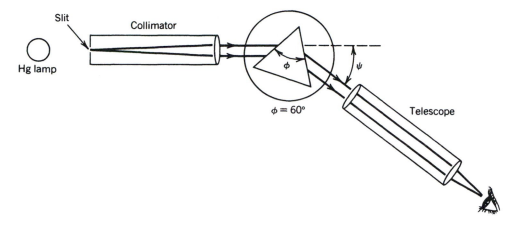

Figure 7

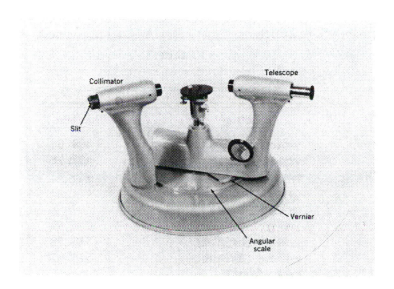

Figure 8

Procedure:

1. Remove the prism and position the telescope so that the crosshairs are centered on the nonmovable edge of the slit. Adjust the telescope eyepiece until the crosshairs and the edge are in sharp focus. Measure the angle γ_0 for undeviated rays.

2. Position the prism on the table and adjust it to the approximate orientation shown in Figure 7.

3. Rotate the telescope and qualitatively observe the spectral lines of mercury. The principal visible wavelengths are given in Table 1. Find the 690.72 nm red line in the telescope. Carefully rotate the prism table to find the angle of minimum deviation for this line. Measure the angle γ for this line. Note that the angle of minimum deviation is

$$\Psi = \mid \gamma - \gamma_0 \mid \qquad (17)$$

See Figure 4. Calculate the index of refraction of the glass prism for the red line.

4.   Ignoring the qualitative observation repeat step 3 for the following wavelengths: λ(yellow) = 579.07 nm, λ(yel-gr) = 546.07 nm, and and λ(blue) = 435.83 nm. Plot n_g versus λ on linear graph paper. Start your y axis at about 1.4 and your x axis at about 400 nm.

Compare your data, n_g versus λ, with those lised in the Handbook of Chemistry and Physics for various types of glass.

Question 5

From your comparison, what kind of glass is your prism made from?

Table 1

Visible Wavelengths of Mercury

Color	λ (nm)
violet (bright)	404.66
violet (faint)	407.78
blue (bright)	435.83
green (moderate)	491.61
green (faint)	--
yel-gr (bright)	546.07
yellow (bright)	576.96
yellow (bright)	579.07
red (faint)	--
red (faint)	--
red (bright)	690.72

23
GEOMETRICAL OPTICS

Apparatus

Optical bench, two +10-cm focal length lenses, +50-cm focal length lens, -15-cm focal length lens, transparent glass sphere (dime store marble), gooseneck lamp, optical bench lamp, screen, objects: transparent aperture, wire mesh.

Introduction

A lens is an image forming device. One of the simplest ways to describe the passage of light through a lens and the formation of an image is by ray optics. Ray optics assumes a beam of light is made up of many light rays that obey Snell's law at each boundary.

The optical axis of a lens system passes through the center of the lens(es). Paraxial rays are rays that make small angles (~10° or less) with the optical axis and lie close to the axis throughout the distance from object to image. For such rays, geometrical arguments can be used to derive formulas for a thin lens. The thin lens equation is an example of such a formula:

$$\frac{1}{o} + \frac{1}{i} = \frac{1}{f} \tag{1}$$

where o is the object distance, i the image distance, and f the focal length of the lens. The equation for the lateral or linear magnification m of a thin lens is another example:

$$m = -\frac{i}{o} \tag{2}$$

Figure 1a shows the formation of a <u>real image</u> by a converging lens. If the object distance is less than the focal length, then the image is <u>virtual</u> as shown in Figure 1b. Some applications of a converging lens for various object distances o are given in Table 1. The corresponding image distances i follow from Equation (1) and the magnification m follows from Equation (2). Negative values of m means the image is inverted |m| > 1 implies it is enlarged.

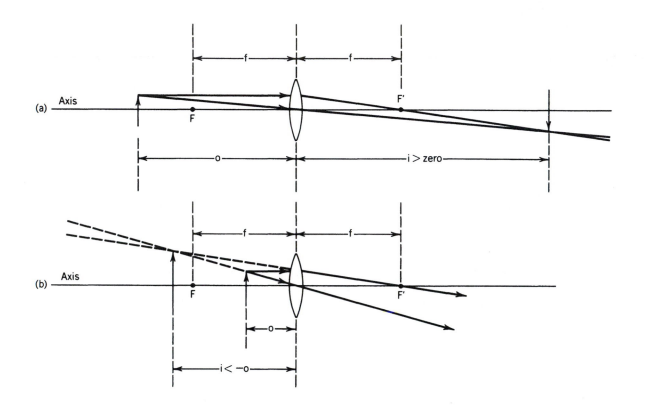

Figure 1

			Table 1	

			Converging Lens	
o	i	m		Application
∞	f	zero		Solar furnace
> 2f	< 2f	-1 < m < zero		Telescope objective lens
2f	2f	-1		Inverter lens in a terrestrial telescope
f < o < 2f	> 2f	< -1		Slide projector lens
f	∞	- ∞		Collimator to form parallel rays
< f	< -o	> 1		Magnifying glass, eyepiece lens

A diverging lens makes rays diverge and by itself it can only form virtual images of real objects. Rays are drawn for two cases in Figure 2.

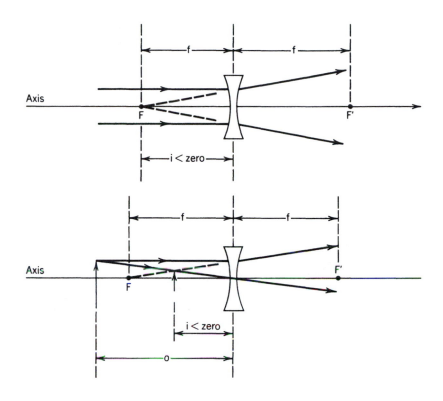

Figure 2

So far we have discussed the ideal case, that is, paraxial rays, where each object point is brought to a focus at a corresponding image point. Departures from the ideal case give rise to slight defects of the image known as <u>aberrations</u>. Three types of aberrations will be discussed.

<u>Spherical aberration</u> is a property of spherical lens. Figure 3 shows a wide beam of rays (hence all are not paraxial rays) incident on a converging lens. The rays are not brought to a focus at a unique point. The resulting image defect is known as spherical aberration. A measure of spherical aberration is $\Delta f/f$ where f is the focal length for paraxial rays. Spherical aberration can be eliminated for a single lens by aspherizing. This is a tedious hand-polishing process by which various regions of one or both lens surfaces are given different curvatures.

<u>Chromatic aberration</u> occurs because of the dispersive properties of transparent media; that is, the index of refraction of such media varies with color. A single lens forms not one image but a series of images, one for each color of light present in the incident beam. This is shown in Figure 4. $\Delta f/f$ is a measure of chromatic aberration, where f is the focal length for yellow light.

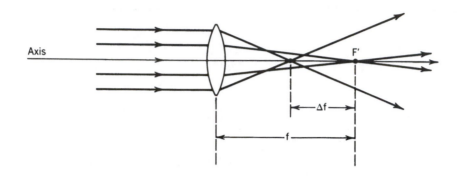

Figure 3

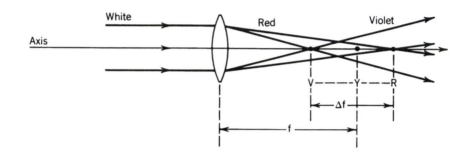

Figure 4

Astigmatic aberration occurs when the object lies some distance from the optic axis; therefore, the rays incident on the lens make an appreciable angle with the optic axis. The images consist of two focal lines, one horizontal and one vertical, and a "circle of least confusion" which is located between the focal lines. This is shown in Figure 5. For clarity rays are not drawn in Figure 5. Note that the focal lines and the circle of least confusion do not lie along the optical axis. The circle of least confusion is a circular image spot and the best image lies in the plane of the spot. As before, $\Delta f/f$ is a measure of astigmatic aberration.

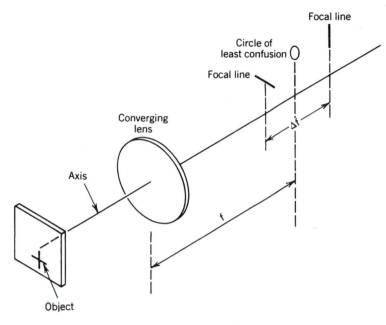

Figure 5

Outcomes

After you have finished the activities in this experiment you will have:

a. determined the approximate focal length of several lenses and a sphere.

b. observed astigmatic and spherical aberrations and determined a measure of each.

c. constructed a microscope and calculated the total magnification.

d. constructed an astronomical telescope.

Focal Lengths

If the object distance o is large compared to the focal length f, then from Equation (1) the focal length is approximately equal to the image distance i. Using the overhead lights, a window, or the sun as the object, determine the approximate focal length of each converging lens and the glass sphere.

Hold the short focal length converging lens (+10 cm is the suggested value) against the diverging lens and form an image of your chosen object. Determine the approximate focal length of the combined lenses. The theoretical focal length f of the combined lenses is

$$\frac{1}{f} = \frac{1}{f_c} + \frac{1}{f_d}$$ (3)

where f_c and f_d are the focal lengths of the converging and diverging lens. Knowing the approximate focal lengths of the converging lens and of the combined lenses, then determine the approximate focal length of the diverging lens.

Question 1

For each lens, what is the percent discrepancy between your measured focal length and the manufacturer's value?

If you or your partner wear spectacle lenses try combining the +10-cm lens or the −15-cm lens (suggested values) with a spectacle lens to form an image. Determine the approximate focal length of each spectacle lens.

Place the glass sphere on a printed page. Observe the image of the print (viewing downward) as you slowly lift the sphere. Note the changes in the image. Does it make sense to you? Determine its approximate focal length. For an index of refraction of 2, parallel light is focussed exactly at the opposite surface of a sphere; if that surface is reflective, then the sphere reflects light back in the direction it came from. This principle is the basis for various reflective coatings (Scotchlite). It also explains the bright reflections of car headlights sometimes seen from an animal's eye at night.

Astigmatic Aberration

The suggested object is a vertical and a horizontal line made with a soft pencil on a 1-cm transparent aperture. Mount the object, +10-cm lens, and the screen on the optical bench as shown in Figure 6; the object is <u>not</u> on the optic axis. Vary the screen position, observing each focal line and the circle of least confusion. See Figure 5. Measure Δf and f; then calculate $\Delta f / f$, which is a measure of astigmatic aberration.

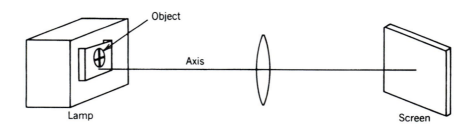

Figure 6

Spherical Aberration

Position the object in Figure 6 so that it is on the optical axis. Cut a paper disk which has a diameter about half that of the +10-cm lens. Use a small piece of tape to mount the disk at the center of the lens. Obtain a sharp image on the screen and measure the focal length. The image is formed from peripheral rays (non-paraxial rays).

Now cut a paper "ring," mount it on the lens, obtain a sharp image, and measure the focal length. In this case the image is formed from paraxial rays. Calculate $\Delta f / f$, where f is the focal length for paraxial rays.

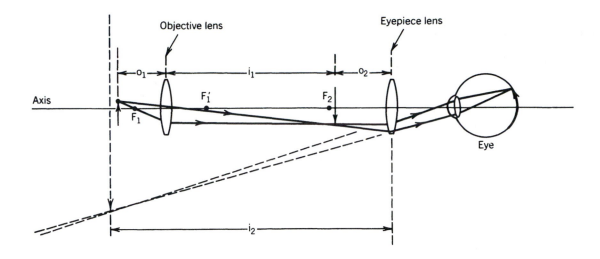

Figure 7

Microscope

A compound microscope has two stages of magnification. First the objective lens produces a magnified real image of the object with $o_1 \geq f_1$. See Figure 7. Then a magnifying glass, the eyepiece lens, is used to further magnify this real image.

The linear magnification M_1 of the objective lens is $-i_1/o_1$, and the angular magnification M_2 of the eyepiece lens is 25 cm/o_2. The total magnification M is

$$M = M_1 M_2 = -\frac{i_1 \times 25 \text{ cm}}{o_1 o_2} \tag{4}$$

The total magnification is the product of M_1 and M_2 because the eyepiece uses the image of the objective as its object.

In principle, any magnification may be obtained with any two converging lenses; in practice the magnification is severely limited by aberrations and geometrical considerations.

Mount two +10-cm lenses and an object on the optical bench. The suggested object is the wire mesh, and it is recommended that the object be illuminated from the front with the gooseneck lamp. Adjust the position of the objective lens until the object is in focus. Move your eye close to the eyepiece lens, then slowly move back. At one position the field of view seems to just fill the eyepiece lens. This is the optimum position for the eye, the "exit pupil." Position the two lenses to obtain a sharply focused and magnified image.

Question 2

What is the calculated value of the total magnification M of your microscope?

Astronomical Telescope: Qualitative

The astronomical telescope is shown in Figure 8. Using the +50-cm focal length lens as the objective and the +10-cm focal length lens as the eyepiece, mount the lenses on your optical bench. Vary the eyepiece position until you obtain a sharp image of a distant object.

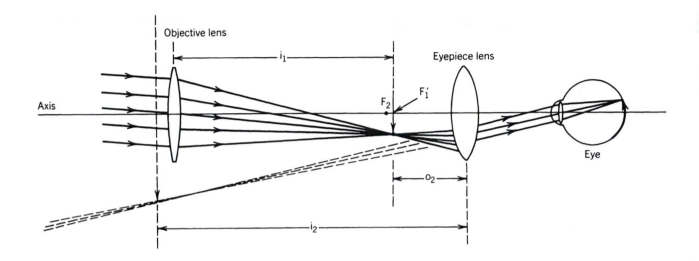

Figure 8

24

DIFFRACTION AND RESOLUTION. BABINET'S PRINCIPLE

Apparatus

Optical bench, laser, lab jack, variable width single slit, double slit, transmission grating, lens of focal length +5 cm, screen, card with 1-cm hole. adhesive metal tape, pin, glass slide with blood smear (usually available from student health center), 30-cm rule. For entire lab: traveling microscopes.

Introduction

A. SINGLE SLIT

Diffraction occurs when a wave passes through a slit. We apply Huygen's principle to the wavefront in the slit, dividing it into infinitesimal coherent radiators. Waves from the coherent radiators then interfere according to the superposition position.

Fraunhofer diffraction occurs when the waves incident on the slit and the detector (screen) are plane waves. This requires either the source and the detector are both located an infinite distance from the slit or lenses are used as shown in Figure 1a. Also if the slit and detector dimensions are very small compared to the curvature of incident spherical wavefronts, then Fraunhofer diffraction occurs to a very good approximation.

Fresnel diffraction occurs when the waves incident on the slit or detector are spherical waves. This is shown in Figure 1b. In this experiment we consider only Fraunhofer diffraction.

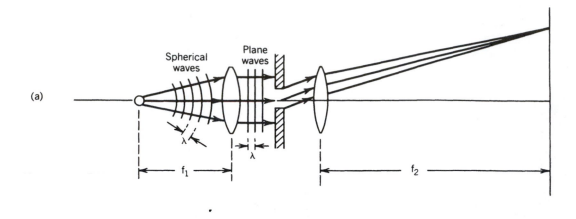

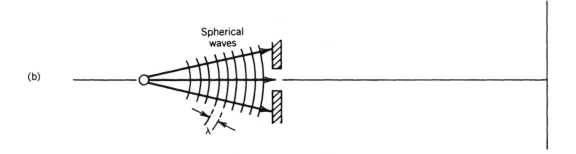

Figure 1

Figure 2 shows a plane wave of wavelength λ incident on a slit of width a. The diffraction pattern, intensity versus θ, is sketched in the figure.

The total electric field (superposition of the fields from each radiator) at a point on the screen specified by r and θ is

$$E(r,\theta,t) = E_0 \frac{\sin \alpha}{\alpha} \cos (kr - \omega t) \qquad (1)$$

where E_0 is the electric field amplitude, k is the wavevector $2\pi/\lambda$, ω is the angular frequency $2\pi/T$, and

$$\alpha = \frac{\pi a}{\lambda} \sin \theta \qquad (2)$$

The intensity is proportional to the square of the total wave, and the time-averaged intensity (for fixed r) $\overline{I}(\theta)$ is:

$$\overline{I}(\theta) \propto \frac{1}{T} \int_0^T E^2 (\theta,t) \, dt \qquad (3)$$

$\bar{I}(\theta)$ is given by

$$\bar{I}(\theta) = I_0 \left(\frac{\sin \alpha}{\alpha}\right)^2 \qquad (4)$$

The angular dependence is contained in α and the function $(\sin \alpha/\alpha)^2$ is called the "diffraction factor" or the "diffraction envelope."

The minima in the diffraction pattern occur where $\bar{I}(\theta) = 0$. This condition requires:

$$a \sin \theta_m = m\lambda \qquad m = 1,2,3,\ldots \qquad (5)$$

where

        m – order number in the diffraction pattern.

        θ_m – angle from the pattern middle to the mth <u>minimum</u> of the diffraction pattern.

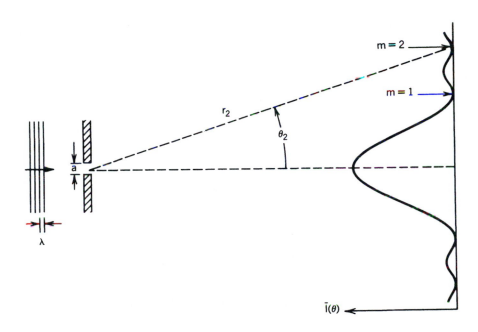

Figure 2

B. DOUBLE SLIT

In the case of two or more slits: (a) the single-slit diffraction patterns (one from each slit) coincide because parallel rays are focused at the same spot, (b) interference takes place inside the "diffraction envelope" corresponding to single-slit diffraction.

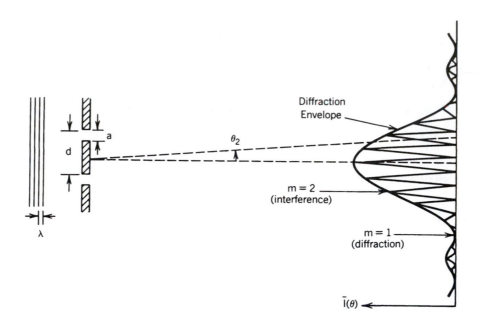

Figure 3

Figure 3 shows a plane wave incident on a double slit, the slits having width a and separation d. $\bar{I}(\theta)$ versus θ is sketched in the figure. Each slit produces a wave given by Equation (1). These two waves superpose to produce a total wave:

$$E(r,\theta,t) = 2E_0 \cos \beta \frac{\sin \alpha}{\alpha} \cos (kr - \omega t) \tag{6}$$

where

$$\beta = \frac{\pi d}{\lambda} \sin \theta \tag{7}$$

and α is given by Equation (2). The time-averaged intensity (for fixed r) is

$$\bar{I}(\theta) = 4I_0 (\cos \beta)^2 (\frac{\sin \alpha}{\alpha})^2 \tag{8}$$

The term $(\cos \beta)^2$ is called the "interference factor." Note that $\bar{I}(\theta)$ involves a product of the "interference factor" and the "diffraction factor."

The interference factor is a maximum for those values of θ which satisfy:

$$d \sin \theta_m = m\lambda \qquad m = 0,1,2,3... \tag{9}$$

where

θ_m - angle from the pattern middle to the mth <u>maximum</u> of the diffraction pattern.

Note in the $\bar{I}(\theta)$ versus θ sketch in Figure 3 that the 5th order maximum is "missing." It is missing because the "diffraction factor" is zero at that angular position. Is it clear that the $\bar{I}(\theta)$ versus θ sketch in Figure 3 is a sketch of Equation (8)?

C. GRATING

A typical grating contains 10,000 slits over a width of about 1 inch or 2.54 cm. Hence the total electric field at a point on the screen is the superposition of 10,000 fields given by Equation (1). The time-averaged intensity for a grating having N slits is:

$$\bar{I}(\theta) = I_0 \left(\frac{\sin N\beta}{\sin \beta}\right)^2 \left(\frac{\sin \alpha}{\alpha}\right)^2 \tag{10}$$

where α and β are given in Equations (2) and (7). Show that Equation (10) reduces to Equation (8) for N = 2. The term $(\sin N\beta/\sin \beta)^2$ is the "interference factor" for a grating having N slits. As for the double slit $\bar{I}(\theta)$ is proportional to the product of an "interference factor" and a "diffraction factor." As N increases, each principal maximum increases in intensity and decreases in width. The angular positions of the principal maxima satisfy

$$d \sin \theta_m = m\lambda \qquad m = 0, 1, 2, 3... \tag{11}$$

where d is the spacing between adjacent slits. The geometry for the grating diffraction pattern is shown in Figure 4.

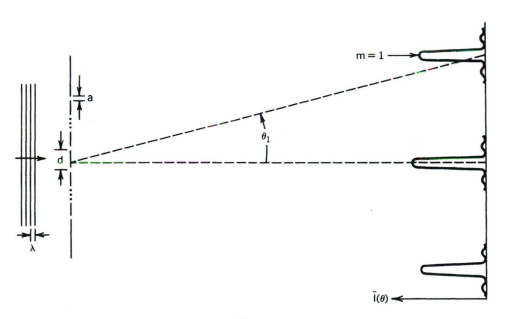

Figure 4

D. CIRCULAR APERTURE

The diffraction pattern of a circular aperture is circular. The equations for the total electric field and the intensity are too advanced for a beginning physics class. The equation which specifies the angular position of the <u>minima</u> has the same form as that for the linear diffraction pattern produced by a single slit:

$$d \sin \theta_m = m\lambda \qquad (12)$$

where d is the hole diameter. However, m is not an integer but is $1/\pi$ times the value at which the first-order Bessel function goes to zero. That is, m = 1.220, 2.233, 3.238, 4.241, ... Thus the first dark ring occurs at an angular displacement given by

$$\sin \theta_1 = \frac{1.22\,\lambda}{d} \qquad (13)$$

This angle is shown in Figure 5.

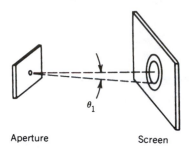

Aperture   Screen

Figure 5

E. BABINET'S PRINCIPLE

Babinet's principle states that the diffraction patterns produced by two complementary screens are the same. The term complementary here signifies that the opaque spaces in one screen are replaced by transparent spaces in the other, and vice versa. For example, a single slit and a thin wire or hair are complementary screens and a circular hole and a disk are complementary screens. These two examples are shown in Figure 6.

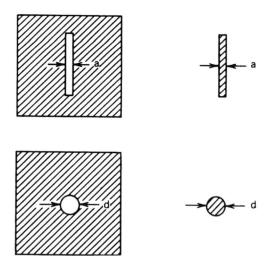

Figure 6

Outcomes

After you have finished the activities in this experiment you will have:

a. observed Fraunhofer diffraction and interference for a single slit, a double slit, and a grating.

b. observed Fraunhofer diffraction for a single circular hole.

c. observed Fraunhofer diffraction and resolution for two circular holes.

d. observed Fraunhofer diffraction for a human hair and red blood cells, and used Babinet's principle to determine the thickness of the hair and the diameter of a red blood cell.

Single Slit, Double Slit, Grating; Qualitative

The laser emits a beam of light whose diameter is about 1 mm. You may enlarge the diameter of the laser beam by passing it through a lens. Place the lens at one end of your optical bench and a screen at the other end. Place the single slit near the lens and illuminate the slit with the enlarged laser beam as shown in Figure 7. Observe the diffraction pattern as you vary the slit width.

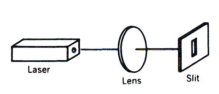

Figure 7

What diffraction pattern would you expect if the slit was reduced to a small square? Test your intuition by reducing the aperture to a small square and then observe the diffraction pattern. Sketch the observed pattern in your lab notebook.

Question 1

What is the equation, analogous to Equation (4), which describes your observed diffraction pattern? Hint: You only need to multiply Equation (4) by the appropriate angular dependent function which describes the vertical diffraction pattern.

Replace the single slit with a double slit and observe the diffraction pattern.

Question 2

What order in the interference maxima corresponding to the first order in the diffraction minima is missing?

Question 3

Using the result from Question 2, what is the ratio of slit width a to slit separation d?

Replace the double slit with a grating and observe the diffraction pattern. Measure two distances such that the angle θ_1 (angular position of the first-order maximum) can be calculated. Then knowing the laser wavelength (your instructor will provide it), calibrate your grating by using Equation (11) to calculate d. Manufacturers usually specify the number of lines per inch, 1/d.

Question 4

What is the percent discrepancy between your value for the number of lines per inch and the value specified by the manufacturer?

Remove the grating from the optical bench, hold it in front of one eye and observe an incandescent lamp through the grating. Discuss the observed diffraction pattern with your lab partner. Does the pattern agree qualitatively with Equation (12)?

Diffraction and Resolution; Qualitative

Stick a piece of adhesive metal tape across the 1-cm hole in the card. Mount the card, screen, and lens on the optical bench as shown in Figure 8. Twirl a pin to make a good circular hole in the tape at the center of the laser spot. Observe the diffraction pattern due to the circular hole.

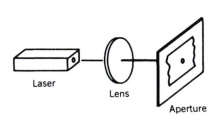
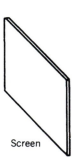

Figure 8

Question 5

How can you tell from the observed pattern of diffraction whether or not the hole is truly circular?

Twirl a second hole about 1 or 2 mm from the first hole. Position the laser so that both holes are illuminated. Observe the diffraction patterns by initially holding the screen a few centimeters from the holes and then slowly move the screen to a few meters away from the holes.

At first you should have no difficulty observing two well resolved spots of light on the screen. As the screen is moved away each spot takes on the character of a diffraction pattern and eventually the two patterns merge into one. Resolution of the spots has been washed out by diffraction.

Question 6

The Rayleigh criterion for resolution stipulates that two images are just resolved when the center of one pattern falls on the first dark band of the other. Is this what you observed? Sketch the diffraction pattern you observe when the Rayleigh criterion is satisfied.

Human Hair Diffraction Pattern

Babinet's principle states that a human hair and a single slit are complementary screens.

Remove the adhesive metal tape from the card having a 1-cm hole. Stretch a hair across the hole and tape it on each side. Mount the card and a screen on the optical bench and illuminate the hair with the laser as shown in Figure 9. The lens, to enlarge the beam diameter, is not needed.

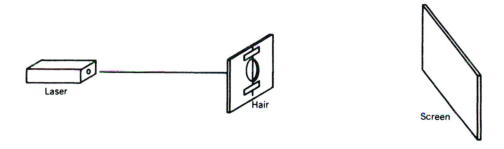

Figure 9

From the geometry of your experiment you can determine the angular position θ_2 of the second minimum. Do so and then calculate the hair thickness using Equation (5). Measure the hair thickness with a traveling microscope.

Question 7

What is the percent discrepancy between the two values for the hair thickness?

Blood Cell Diffraction Pattern

A red blood cell and a circular hole are complementary screens. Mount a glass slide which has a blood smear and a screen on the optical bench. Place the screen about 6 or 8 cm from the slide. Adjust the lateral position of the laser until a circular diffraction pattern is observed on the screen. Shining the laser beam on the edge of the blood smear (where the smear is thinner) will produce the best diffraction pattern.

You can determine the angular position θ_1 of the first dark ring from the geometry of your experiment. Do so and then calculate the diameter d of the red blood cell.

Question 8

What is your calculated value of the diameter of a red blood cell?

The diameter of a red blood cell does depend on its age (its mean lifetime is about 128 days) and an average value is 7.5×10^{-6} m.

25
POLARIZATION

Apparatus

Optical bench, lamp, three polaroids, miscellaneous materials (perhaps glass, plastic, cellophane, etc.), glass plate with protractor and sighting arm, quarter-wave plate, goose neck lamp. Optional: Karo syrup and distilled water solution.

Introduction

An em wave has electric and magnetic fields which are perpendicular to each other and to the direction of propagation. We may describe the direction of the electric field $\vec{E}$ in an em wave at one point in space and at one time by its components E_x and E_y as shown in Figure 1, where it is assumed the wave propagates along the z axis. The direction of propagation is specified by the wavevector $\vec{k}$. For a monochromatic wave, the components of $\vec{E}$ are given by

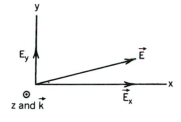

Figure 1

$$E_x(z,t) = E_{x,0} \sin(kz - \omega t) \qquad (1)$$

$$E_y(z,t) = E_{y,0} \sin(kz - \omega t + \phi) \qquad (2)$$

where the amplitudes $E_{x,0}$ and $E_{y,0}$ may be any size. The phase difference ϕ represents a time delay in the waves E_x and E_y at the given point in space, as shown in Figure 2.

In unpolarized light, the phase angle ϕ is randomly fluctuating. Light from the sun, incandescent lamps, and fluorescent lamps is unpolarized.

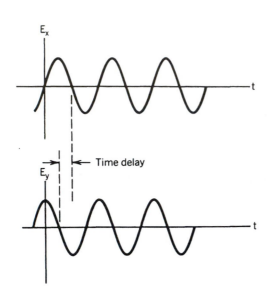

Figure 2

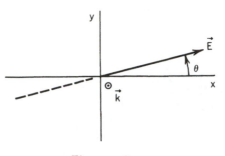

Figure 3

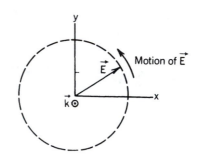

Figure 4

Polarized light has a constant phase angle. Some special cases are:

(1) In linearly or plane polarized light, the phase angle ϕ is zero; hence E_x and E_y are in phase and the $\vec{E}$ field vector oscillates along the line which makes an angle θ with the x-axis. See Figure 3. The angle θ is determined by the amplitudes $E_{x,0}$ and $E_{y,0}$:

$$\tan \theta = \frac{E_{y,0}}{E_{x,0}} \qquad (3)$$

(2) In right-hand circularly polarized light the phase angle is $+\pi/2$ and $E_{x,0} = E_{y,0} = E_0$. In this case Equation (1) and (2) become

$$E_x(z,t) = E_0 \sin (kz - \omega t) \qquad (4)$$

$$E_y(z,t) = E_0 \sin (kz - \omega t) \qquad (5)$$

The $\vec{E}$ field motion in the plane of the page is shown in Figure 4. With the right thumb along $\vec{k}$, the fingers curl in the direction that the tip of the $\vec{E}$ vector travels; hence, the name right-hand circularly polarized light. The $\vec{E}$ vector in space at a given time describes a right-hand screw helix as shown in Figure 5.

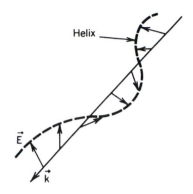

Figure 5

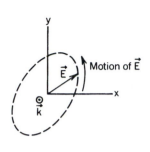

Figure 6

Left-hand circularly polarized light occurs when $\phi = -\pi/2$ with $E_{x,0} = E_{y,0}$. In this case with the left thumb along $\vec{k}$ the fingers curl in the direction that the tip of the $\vec{E}$ vector travels.

(3) In right-hand elliptically polarized light $\phi = +\pi/2$ and $E_{x,0} \neq E_{y,0}$ or when $\phi \neq \pm\pi/2$. The E field motion in the plane of the page is shown in Figure 6, where the tip of the $\vec{E}$ vector traces out an ellipse in the "right-hand" sense.

In left-hand elliptically polarized light $\phi = -\pi/2$ and $E_{x,0} \neq E_{y,0}$.

Outcomes

After you have finished the activities in this experiment you will have studied polarization produced by absorption, reflection, scattering, and refraction.

Polarization by Absorption

Polaroid is the tradename of a material which produces polarized light by absorption. It consists of polyvinyl alcohol, a rubber-like material, which has been subjected to a large tensile strain and mounted in plastic. This orients the polyvinyl alcohol molecules parallel to the direction of the strain. The $\vec{E}$ field parallel to the strained, long molecules causes the molecular electrons to oscillate; hence, the component of the $\vec{E}$ field parallel to the strain direction is absorbed. The arrow shown on the polarizer (sheet of polaroid) in Figure 7 specifies the polarizing direction, which is perpendicular to the direction of strain. The $\vec{E}$ field in the polarizing direction passes through with only slight absorption.

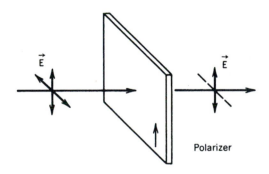

Figure 7

Place a lamp and a polarizer on the optical bench.   Using your eye as the detector, observe the lamp through the polarizer as you rotate it.

Question 1

Does the lamp emit polarized light?   Briefly explain your answer, based on your observation.

Place a second polarizer on the optical bench.   Leaving the orientation of the polarizer which is closer to the lamp fixed, rotate the other one as you observe the lamp. You should find that when the two polarizers are "crossed" the lamp is nearly extinguished.

"Cross" the two polarizers and then insert a third polarizer between them.   Rotate the third polarizer until you observe the light to be the most intense.   For this case draw a diagram showing the orientation of the polarizing direction for each polaroid.

Question 2

For the orientation of the three polarizers shown in your diagram, explain why light is transmitted.

Remove the third polarizer, leaving the other two crossed.   Place miscellaneous materials (plastic, glass, cellophane, etc. may be available) between the crossed polarizers. (This technique is used to examine stress in glass and plastic models of tools.)

Polarization by Reflection

When unpolarized light is incident at an angle ϕ_p (Brewster's angle) which, satisfies

$$\tan \phi_p = \frac{n_2}{n_1} \qquad (6)$$

then the reflected light is polarized as shown in Figure 8.

The suggested apparatus to observe reflected light is shown in Figure 9. At Brewster's angle you will be able to extinguish the reflected light by a suitable rotation of the polarizer.   For a light source use the lamp at your station or the overhead lights; an extended sense is better than a point source here.

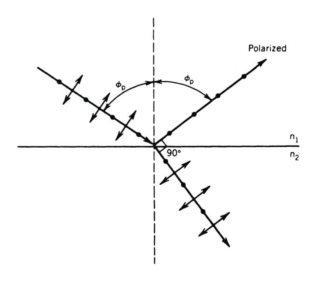

Figure 8

Question 3

From your observation what is the index of refraction of the glass plate?

Viewing through the polarizer look at light reflected from any surfaces in the room or outside. Is the light partially or totally polarized? Light reflected from dielectrics tends to be polarized (the effect is smaller for metallic reflection). Polaroid sunglasses make use of this in reducing reflected glare.

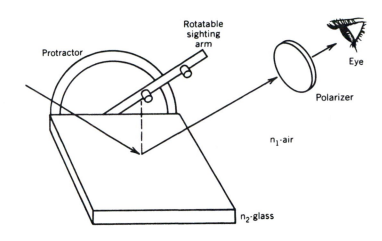

Figure 9

Polarization by Scattering

The electric field of light striking an atom causes the atomic electrons to oscillate parallel to the electric field, and the oscillating charges radiate or scatter light. The electric field of the scattered light is, among other things, proportional to the acceleration of the electrons; hence, the $\vec{E}$ field is proportional to the displacement of the electrons.

In Figure 10 unpolarized light is incident on an atomic electron, causing oscillation of the electron in the xy-plane. If you observe the oscillating charge from any position in the xy-plane, you observe "vertically" polarized light.

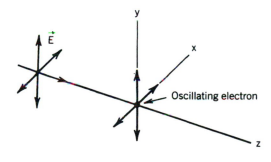

Figure 10

Question 4

In order to observe scattered sunlight which is linearly polarized, what direction should you look?

Draw a diagram in your notebook showing light from the sun incident on the sky and the direction you look.

Test your answer to Question 4 by looking out a window or going outside and viewing the sky through a polarizer.

Polarization by Refraction

Glass is an example of a material which is normally <u>optically isotropic</u>, that is, the index of refraction does not depend on the polarization or the direction of propagation of light.

Calcite is <u>optically anisotropic</u>, that is, the index of refraction depends on the polarization and the direction of propagation of light. Figure 11a shows unpolarized light incident at some angle to the optic axis of an optically anisotropic crystal. The unpolarized light splits into two beams which are polarized at right angles to each other. The <u>ordinary</u> or <u>o-beam</u> obeys Snell's law and the <u>extraordinary</u> or <u>e-beam</u> does not. The velocity of the o-beam is

$$v_o = \frac{c}{n_o} \qquad (7)$$

and that of the e-beam is

$$v_e = \frac{c}{n_e} \qquad (8)$$

where n_o and n_e are the principle indices of refraction and $n_o \geq n_e$ or $v_o \geq v_e$.

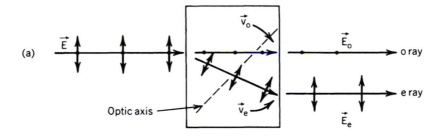

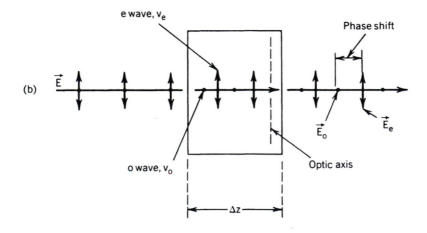

Figure 11

Figure 11b shows unpolarized light which is incident normal to the optic axis. For this orientation $v_e > v_o$ and the beams emerge having a phase difference ϕ, where ϕ is proportional to the crystal thickness Δz. If the crystal thickness is chosen so that $\phi = 90°$, then the crystal is called a <u>quarter-wave plate</u>.

A quarter-wave plate combined with a linear polarizer, as shown in Figure 12, produces circularly polarized light.

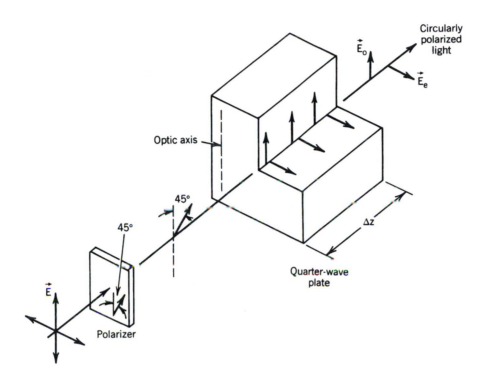

Figure 12

Question 5

Why must the $\vec{E}$ field of the linearly polarized light impinge on the quarter-wave plate at an angle of 45° to the optic axis?

Question 6

If the angle between the optic axis and the $\vec{E}$ field of the incident polarized light is not 45° but is 60°, say, then would the light emerging from the quarter-wave plate be polarized? What kind of polarized light would emerge?

Place two polarizers on the optical bench and cross them. Insert a quarter-wave plate between the cross polarizers. Rotate the quarter-wave plate until you achieve circularly polarized light emerging from the quarter-wave plate.

Question 7

How will you determine when the light is circularly polarized?

Optical Activity (Optional)

A remarkable effect of polarization is observed in materials composed of molecules which do not have reflection symmetry, i.e., molecules which are not invariant under reflection across some plane through the molecule. Such molecules can exist in "left-handed" and "right-handed" forms, identical in composition and physical properties, but differing optically. They may exhibit optical activity, whereby a linearly polarized light passes through the material, the direction of polarization is rotated through some angle θ as shown in Figure 13.

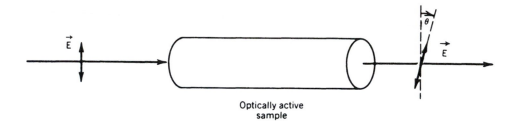

Figure 13

Glucose, $C_6H_{12}O_6$, occurs in two optically different forms: (1) dextros; dextrorotatory glucose, or D-glucose rotates the plane of polarized light in a clockwise direction as seen by the observer in the diagram, (2) levorotatory glucose or L-glucose rotates the plane of polarized light in a counterclockwise direction as seen by the observer shown above.

On your optical bench set up the polarimeter shown in Figure 14. Measure the rotation, θ, of the plane of polarization produced by the optically active glucose solution [Karo syrup (pure D-glucose) and distilled water]. You can do this visually by observing the angular position of the analyzer polaroid which produces zero intensity with and without the sample present.

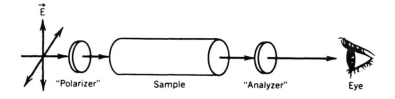

Figure 14

The rotation of a given solution, θ, depends on the number of chiral molecules that the light beam encounters. For a fixed concentration of chiral molecules, C, the longer the sample tube, ℓ, the more the plane of polarization is rotated because more molecules are encountered over a longer path. Thus, $\theta \propto \ell$. If the tube length is constant, then the higher the concentration of chiral molecules, C, the greater is the rotation, θ. Thus $\theta \propto C$, hence $\theta \propto \ell C$ or $\theta = $ constant $\times \ell C$ where the constant is defined to be the specific rotation, $[\alpha]_\lambda^T$, which is the number of degrees of rotation per unit concentration per unit path length. By definition

$$[\alpha]_\lambda^T = \frac{\theta}{C\ell}$$

where

      C = concentration in grams per ml of solution,

      ℓ = length of the light path in the solution, measured in decimeters (1 dm = 10 cm),

      θ = observed rotation in degrees,

      λ = wavelength of light,

      T = temperature of the solution in degrees celsius.

Knowing $[\alpha]_{\lambda}^{T}$ and ℓ, C may be calculated once θ has been measured; such measurements are used in wine production.

26

ELECTRON ORBITS IN A MAGNETIC FIELD. e/m

Apparatus

Fine beam tube with auxiliary equipment.

Introduction

The fine beam tube is a special type of cathode ray tube which is used to show the track of a fine electron beam and the behavior of the beam in magnetic fields. A simplified diagram of the apparatus is shown in Figure 1. The beam emerges from the hole in the anode with a speed v determined by the accelerating potential V_a:

$$\frac{1}{2} mv^2 = eV_a \tag{1}$$

where m is the mass of the electron and e is the magnitude of its charge. The beam enters the magnetic field B produced by the Helmholtz coils. The magnetic field at the center is given by

$$B = \mu_0 \left(\frac{4}{5}\right)^{\frac{3}{2}} \frac{nI}{R} \tag{2}$$

where $\mu_0 = 4\pi \cdot 10^{-7}$ N/A^2, n = number of turns on each coil, R = coil radius, and I is the coil current. Your instructor will provide values for n and R.

Newton's second law applied to a single electron in the magnetic field is

$$evB \sin\theta = m \frac{(v \sin\theta)^2}{r} \tag{3}$$

where r is the orbit radius and θ is the angle between $\vec{v}$ and $\vec{B}$. The fine beam tube may be rotated relative to the $\vec{B}$ field, thus changing the angle θ.

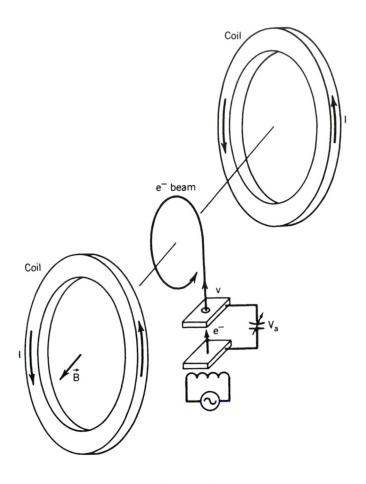

Figure 1

The track of the electron beam is visible because the tube contains a gas (often hydrogen or neon) at a pressure of approximately 1.33 pascals (10^{-2} mm Hg). Inelastic collisions of the beam electrons with the atomic electrons results in the excitation of the atomic electrons, and de-excitation occurs with the emission of visible light.

The electron beam is a few millimeters wide, and it is worthwhile to examine the mechanisms which broaden the electron beam. A broadened beam is shown in Figure 2 where r_i and r_o are the inner and outer radii.

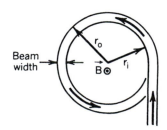

Figure 2

1. _Thermionic emission broadening_. Equation (1) assumes each electron is initially emitted by the cathode with zero velocity. This is not the case. The cathode is heated to approximately 2500 K, hence, the electrons are emitted by the cathode with a distribution of speeds ranging from zero to large values. This effect, called _thermionic emission_, was first discovered shortly after the invention of the electric lamp (thermionic emission from the filament) and it is crucial for all devices requiring electron guns, e.g., cathode ray tubes and electron microscopes.

The distribution of speeds does not follow Maxwell's speed distribution law, and the average kinetic energy $\bar{K}$ of an electron emerging from the cathode is

$$\bar{K} = 2kT \tag{4}$$

rather than 3/2 kT, where k is the Boltzmann constant and T is the absolute temperature. The average speed $\bar{v}$ of an electron emerging from the hole in the anode may be determined from energy conservation:

$$\frac{1}{2}m\bar{v}^2 = eV_a + 2kT \tag{5}$$

Some electrons will emerge with speeds greater than $\bar{v}$ and others will have speeds less than $\bar{v}$.

Question 1

Assuming V_a = 200 V and T = 2500 K, what fraction of the average speed $\bar{v}$ is due to thermionic emission?

To understand how thermionic emission effects the width of the beam we need to examine Equation (3). Suppose the fine beam tube is rotated so that θ is 90°. Then solving Equation (3) for r we find

$$r = \frac{mv}{eB} \tag{6}$$

Thus a distribution of speeds gives rise to a distribution of radii, hence a broadening of the electron beam. Note from Equation (5) that the average speed $\bar{v}$ increases as T increases, hence from Equation (6) the average radius $\bar{r}$ increases as T increases. Thus thermionic emission causes a temperature dependent broadening of the outside edge of the beam. Figure 3 shows a cross section of the beam and general features of beam broadening.

2. _Collisional broadening_. The electrons in the beam collide with the gas atoms and are deflected. This effect, called collisional broadening, is more pronounced for low speed electrons which in general are scattered through larger angles than high speed electrons. This effect would cause the inner edge of the beam to be less distinct than the outer edge. See Figure 3.

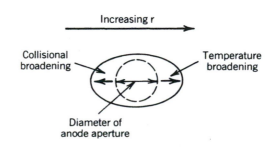

Figure 3

Outcomes

After you have finished the activities in this experiment you will have:

a.   qualitatively observed the radius r of the circular orbits as a function of the magnetic field B and the accelerating potential V_a.

b.   measured the diameter 2r of circular orbits as a function of both the magnetic field B and the accelerating potential V_a.

c.   calculated the ratio of charge/mass for the electron.

Qualitative Observations

In your lab notebook, use Equations (1) and (3) (with $\theta = 90°$) to eliminate the electron velocity v and hence obtain an equation involving r, V_a, B, and constants. Solve this equation for r.   Note that this result is approximately valid for electrons emerging from the cathode with low speeds, that is, thermionic emission effects are ignored.

Before turning on the equipment, identify the following control knobs: accelerating voltage V_a, magnetic field current I, and electron gun filament voltage.

Obtain a beam at an accelerating voltage of 200 volts.   Vary the magnetic field by changing the current I in the coils.   At a fixed current in the coils, vary the accelerating voltage.   DO NOT EXCEED THE RATED MAXIMUM.

Question 2

For a fixed accelerating voltage V_a, how do you expect the orbit radius r to change as you change B?   What do you observe?

Question 3

For a fixed magnetic field B, how do you expect the orbit radius r to change as you change V_a?   What do you observe?

Rotate the tube slightly in its mount.   Observe the helical orbits which result. Reset for orbits in a plane.

Quantitative Observations, e/m

In this experiment there are two independent variables V_a and B, and one dependent variable r.   Since there are two independent variables, r can be measured as a function of the independent variables in several ways and in each way the data may be graphed and e/m determined.   For example, we could set V to a constant value and measure r as a function of B, or we could set B and measure r as a function of V.

Question 4

If V remains constant and r is measured as a function of B, then B plotted versus what function of r should yield a straight line?   What would be the theoretical slope of the line?

Question 5

If B remains constant and r is measured as a function of V, then V plotted versus what function of r would yield a straight line? What would be the theoretical slope in this case?

It is suggested that you choose a value for B and measure 2r for 3 or 4 values of V. Then choose a value for V and measure 2r for 3 or 4 values of B. Plot all of the data on a graph of V/B^2 versus r^2.

Question 6

What would be the theoretical slope of a plot of V/B^2 versus r^2?

Determine $e/m \pm \delta(e/m)$ from your graph. The accepted value is 1.76×10^{11} C/kg.

Question 7

What is the percent discrepancy between your value of e/m and the accepted value?

If the accepted and experimental values do not agree within the precision of your experiment, then identify any systematic errors which may explain this.

27
PHOTOELECTRIC EFFECT. WORK FUNCTION, PLANCK'S CONSTANT

Apparatus

Phototube, mercury arc lamp, DC power supply, interference filters, voltmeter, micro-ammeter, 5000-Ω potentiometer, two 2700-Ω $\frac{1}{2}$-watt resistors.

Introduction

In part A the energy levels of conduction electrons in a metal are discussed. The absorption of photons by a metal and the ejection of electrons are discussed in part B.

A. ELECTRONS IN METAL

Isolated atoms have non-overlapping coulomb potential wells with quantized atomic energy levels. The potential wells of two identical isolated atoms and their hypothetical energy levels are shown in Figure 1a. As the separation of the two atoms decreases the coulomb wells overlap and one or more atomic energy levels become molecular energy levels. In Figure 1b the solid curves give the molecular potential and the dash curves give the potential of each isolated atom. E'_5 and E''_5 are molecular energy levels and E_1, E_2, E_3, and E_4 remain atomic energy levels. An electron occupying the molecular levels E'_5 or E''_5 is not localized on either atomic nucleus, rather it is shared by the two atoms. Electrons occupying $E_1, \ldots, E_4$ remain localized on their original nucleus.

A metal crystal is a three-dimensional array of overlapping coulomb wells where one or more atomic energy states of each atom become "conduction electron" states which extend throughout the metal. A one-dimensional metal formed from N atoms is shown in Figure 1c. The N E_5 energy levels of the isolated atoms become the N conduction electron states; $\varepsilon_1, \varepsilon_2, \ldots, \varepsilon_N$.

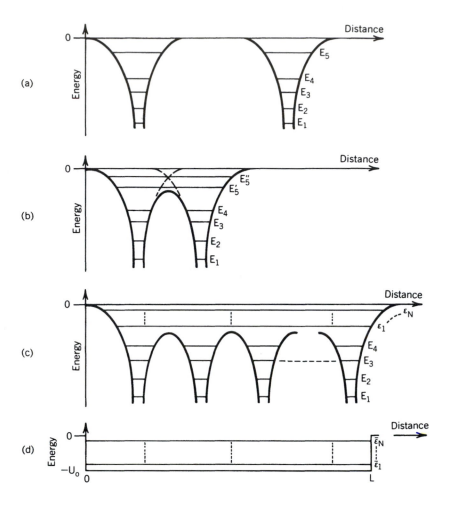

Figure 1

We can determine average energies, $\bar{\epsilon}_1$, $\bar{\epsilon}_2$, ..., $\bar{\epsilon}_N$, of the conduction electrons by ignoring the coulomb wells, i.e., we assume the conduction electrons are in a "square" well of finite depth $-U_0$ as shown in Figure 1d. Assuming each atom contributes a single electron, we now have N electrons free to move in a one-dimensional well of dimension L, where L is the length of the metal crystal. We describe the possible electron states by standing waves Ψ_n, $n = 1, 2, \ldots N$. Since the well has finite depth each wave extends slightly outside the well and we specify an average wavelength $\bar{\lambda}_n$, $n = 1, 2, 3, \ldots, N$. The general wave Ψ_n has an average wavelength $\bar{\lambda}_n$ which depends on L and n, and we take the average wavelength to be

$$\bar{\lambda}_n = \frac{2L}{n} ; \qquad n = 1, 2, \ldots, N \tag{1}$$

Corresponding to each wavelength there is

an average momentum:
$$\bar{p}_n = \frac{h}{\bar{\lambda}_n} = \frac{h}{2L} n \tag{2}$$

an average kinetic energy:
$$\bar{K}_n = \frac{\bar{p}_n^{\,2}}{2m} = \frac{h^2}{8mL^2} n^2 \tag{3}$$

where h is Planck's constant and m is the mass of the electron. Since the potential energy of the electrons is $-U_0$, the total average energy $\bar{\varepsilon}_n$ is

$$\bar{\varepsilon}_n = \bar{K}_n - U_0 = \frac{h^2}{8mL^2} n^2 - U_0 ; \qquad n = 1, 2, \ldots, N \tag{4}$$

Note that $\bar{\varepsilon}_n$ depends on the fundamental constants h and m, and on the dimension L of the crystal.

In Figure 2 the first three energies are specified by the dashed lines and the standing waves are given by the solid lines.

Question 1

For a metal crystal of dimension L = 1 cm, what is the energy separation of $\bar{\varepsilon}_1$ and $\bar{\varepsilon}_2$? What wavelength photon would be emitted or absorbed by an electron making a transition between these two levels?

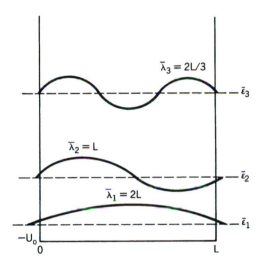

Figure 2

The quantum numbers describing the state of an electron are the energy quantum number n, n = 1,2,...,N, and the spin quantum number, spin up or spin down. The Pauli Exclusion Principle states that no two electrons may have the same quantum numbers; therefore, two electrons, one with spin up and the other with spin down, may occupy each energy level $\bar{\epsilon}_n$. The occupied states are shown in Figure 3, where it is assumed that the electrons occupy the lowest energy states, that there are N electrons, and that N/2 is an even integer. The arrow indicates the spin state. The spacing of the energy levels is not to scale. The minimum energy to remove an electron from the well is W, called the <u>work function</u> of the metal. Table 1 gives values of the work function for several metals.

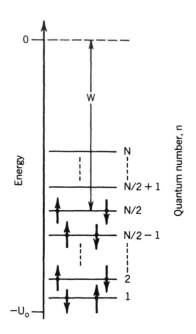

Figure 3

Table 1

Metal	Work Function W(eV)
Cesium	1.9
Beryllium	3.9
Mercury	4.5
Gold	4.8

B. PHOTOELECTRIC ABSORPTION

Prior to 1905 it was widely believed that electromagnetic radiation had <u>wave properties</u> only, i.e., the phenomenon was explainable from the point of view of the interference of waves according to the superposition principle. In 1905 Einstein introduced the concept that, on occasion, radiation has <u>particle properties</u>, i.e., the phenomenon is explainable from the point of view of localized particles (photons), which travel well-defined trajectories to arrive at a specified point at a specified time.

Einstein postulated that the link between a particle property (momentum p) of radiation and a wave property (wavelength λ) is Planck's constant h:

$$P = \frac{h}{\lambda} \tag{5}$$

The total energy E of the particle is related to its momentum p and rest mass energy m_0c^2 by the special theory of relativity:

$$E = \sqrt{p^2c^2 + m_0^2c^4} \tag{6}$$

where c is the speed of light and m_0 is the particle's rest mass, which is zero for the photon. Hence, for the photon

$$E = pc = \frac{h}{\lambda}c = h\nu \tag{7}$$

We presently recognized that the particle properties of radiation are emphasized when the emission or absorption of radiation is studied, and the wave properties are emphasized when the behavior of radiation in moving through a system is studied.

Question 2

For each of the following is it the wave or the particle aspect of radiation that is more prominent: (a) image formation by a lens, (b) Compton scattering, (c) diffraction?

If we shine radiation of an appropriate frequency on a metal, we observe emission of electrons. The net result of the photoelectric effect is that a photon goes in and an electron comes out. It is not too gross an oversimplification to say that the photon knocks out a conduction electron; however, the interaction does not involve a single electron only, since the electron is bound in a potential well.

Figure 4 shows the absorption of a photon having energy $h\nu$ by a single atom and the ejection of an electron at some angle θ. Conservation of momentum requires that the atom recoil; therefore, the entire atom is involved in the interaction and not just a single electron. The entire photon energy is absorbed and an electron ejected with a kinetic energy

$$K = h\nu - B_e \tag{8}$$

where B_e is the binding energy of the ejected electron.

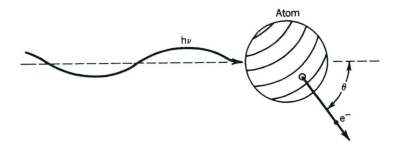

Figure 4

A conduction electron is bound to all of the metallic ions of a crystal and the minimum binding energy is the work function W. An electron occupying a state lower in energy than $\bar{\varepsilon}_{N/2}$ has a larger binding energy. The maximum kinetic energy of the ejected electron is

$$K_{max} = h\nu - W \tag{9}$$

Note that an electron is not ejected if $h\nu < W$.

Question 3

For each metal listed in Table 1, calculate the longest wavelength photon which will eject an electron. In each case specify whether the photon is visible light or UV radiation.

Outcomes

After finishing the activities in this experiment you will have:

a. determined if the phototube is ideal by graphing an I versus V characteristic curve.

b. measured the stopping potential V_0 as a function of photon frequency.

c. determined Planck's constant.

d. determined the work function of the phototube.

Experiment

The suggested circuit is shown in Figure 5.

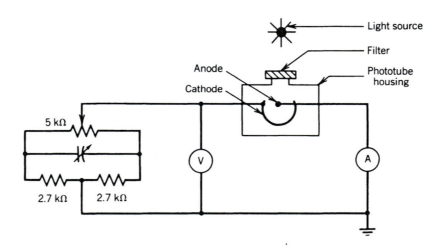

Figure 5

The filter allows the selection of photons of a particular wavelength or frequency. The electrons are ejected from the cathode which is often coated with cesium antimonide. Cesium antimonide has a relatively high quantum efficiency (ratio of electrons emitted to photons absorbed) and it has a low work function (hence, visible light is an acceptable photon source). Quantum efficiency versus wavelength is shown in Figure 6 for two phototubes.

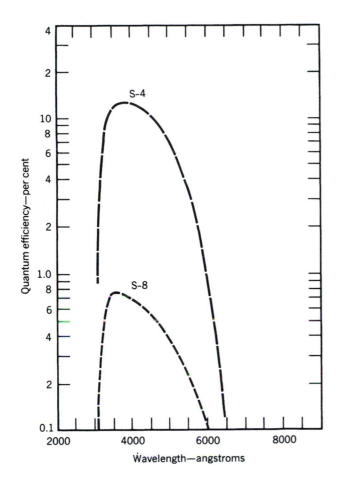

Figure 6

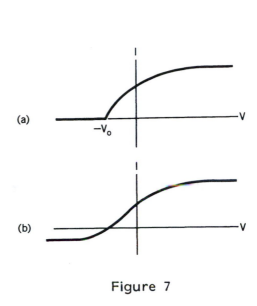

Figure 7

In Figure 5, the 5kΩ potentiometer may be adjusted so that the cathode potential V is positive or negative relative to the anode (ground) and the photocurrent can be measured as a function of the potential across the tube. Figure 7a is a graph of photocurrent I versus potential V across an ideal tube for fixed light intensity and frequency. The potential $-V_0$ prevents the most energetic electrons from reaching the anode, hence the current goes to zero as V approaches $-V_0$. For $V = -V_0$, conservation of energy applied to an electron ejected with the maximum kinetic energy K_{max} yields

$$K_{max} = eV_0 \tag{10}$$

Combining Equations (9) and (10):

$$eV_0 = h\nu - W \tag{11}$$

By measuring V_0 as a function ν, where ν is determined by the filter, the data can be plotted to determine h and W.

First, it is worthwhile determining if your tube is ideal. An I versus V curve for a non-ideal tube is shown in Figure 7b. Such a tube has some photoelectric material (cesium antimonide, e.g.) deposited on the anode. When light strikes the anode, electrons are emitted and then collected at the cathode creating a "reversed" photocurrent. In this case the tube does not have a sharp cutoff of current at a certain potential $-V_0$. It is suggested that a thin strip of opaque tape be attached to the tube to cast a shadow on the anode to shield it from direct light from the mercury lamp. This will reduce the reversed current; however, some scattered light will strike the anode.

Using the green filter (it passes the 546.07-nm line), determine if your tube is ideal by varying V from -3V to +3V in small increments and measuring I for each value of V. Plot V versus I.

The lifetime of the phototube will be increased if you shield it from the lamp when changing filters.

Question 4

Knowing the maximum photocurrent for the 546.07-nm line and using the appropriate efficiency versus wavelength curve given in Figure 6, what is the number of photons per second reaching the cathode?

Measure V_0 several times for each filter and calculate an average value. (Wavelengths of the spectral lines of mercury are listed in Table 1 of Experiment 22.) Plot your data and determine h and W from the graph. See Equation (11).

Question 5

If the tube is not ideal, then does this introduce a systematic or random error in the values for V_0?

Question 6

How would error(s) in V_0 caused by a non-ideal tube affect the values of h and W?

Question 7

What is the percent discrepancy between your value of h and the accepted value (6.626×10^{-34} J·s)?

28

ELECTRON DIFFRACTION. LATTICE CONSTANT

Apparatus

Electron diffraction tube and power supply, 30-cm plastic rule.

Introduction

Prior to 1924 it was widely believed that electrons, protons, golf balls, etc. had particle properties only. In 1924 Louis de Broglie postulated that any particle having a momentum p has associated with it a <u>matter wave</u> of wavelength λ:

$$\lambda = \frac{h}{p} = \frac{h}{mv} \tag{1}$$

where h is Planck's constant, m is the particle's mass, and v is its velocity. De Broglie's postulate, like Einstein's for radiation, drew together apparently unrelated ideas, a wave property λ and a particle property p with Planck's constant h providing the link.

In 1927, Davisson and Germer, working at Bell Labs, experimentally verified de Broglie's hypothesis. The apparatus they used is shown in Figure 1. By changing the angular position of their detector they observed maxima and minima corresponding to the interference of reflected electron waves whose wavelengths were in excellent agreement with de Broglie's hypothesis.

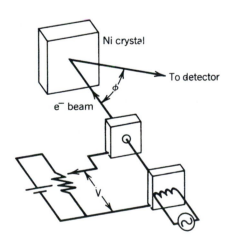

Figure 1

Question 1

Derive the relation between λ and V (accelerating voltage shown in Figure 1) by using Equation (1) and energy conservation. Then evaluate the constants to show:

$$\lambda(\text{Å}) = \sqrt{\frac{150}{V}} \tag{2}$$

where V is in volts.

A crystal has a regular array of evenly spaced atoms and therefore it acts as an excellent diffraction grating for waves of an appropriate wavelength. The waves are reflected from successive planes of atoms in the crystal as shown in Figure 2. The first plane need not coincide with a surface of the crystal or be parallel to a surface.

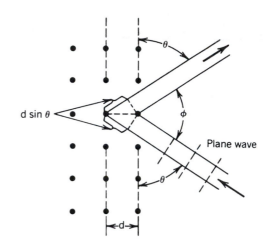

Figure 2

Important point. The interference is not between a wave associated with one electron and a wave associated with another, rather it is interference between different parts of the wave associated with a single electron that have been scattered from two or more planes in the crystal. This can be verified by observing the diffraction pattern as a function of electron beam intensity. A low intensity beam, with one electron going through the apparatus at a time, produces the same pattern as a high intensity beam.

We now consider the geometric law (Bragg's law) for diffraction by a crystal of a monochromatic beam of particles. Bragg's law applies to any species of wave particle, e.g., X-ray photons, electrons, neutrons, etc. The path difference of the two rays shown in Figure 2 is 2d sin θ, where d is the spacing between successive planes of atoms and θ is the angle between the incident beam and the reflection plane. Note in particular: Bragg's law specifies that the reflection is specular, i.e., the angle of reflection equals the angle of incidence, and θ is not the angle to the normal, but to the plane. The two reflected waves will interfere constructively if the path difference is an integral number of wavelengths:

$$2d \sin \theta = n\lambda \tag{3}$$

This result depends on the spacing d of the planes of atoms but is independent of the atomic arrangement within each plane. The intensity of Bragg reflection depends on the surface density of atoms within each plane.

Figure 3 shows some of the many possible sets of parallel planes in a simple square crystal with lattice constant a and the spacing d for each set. It is useful to have a general way to describe a particular set of planes. A set of planes is usually described by their Miller indices, which are established as follows for a square lattice.

Suppose a particular plane of a given set has intercepts qa and ra (q and r integers) with the axes. We could call this plane the qr plane; however, that turns out to be a clumsy notation. The Miller indices of the plane are given by two numbers h and k such that

$$h = \frac{m}{q} \qquad (4)$$

$$k = \frac{m}{r}$$

where m is an integer chosen such that h and k are integers having no common factor > 1. The Miller indices of several individual planes are given in Figure 4. One or both index may be negative when the corresponding intercepts are negative and a negative index is written with a "bar" above it: $(\bar{h}k)$, $(h\bar{k})$, or $(\bar{h}\bar{k})$.

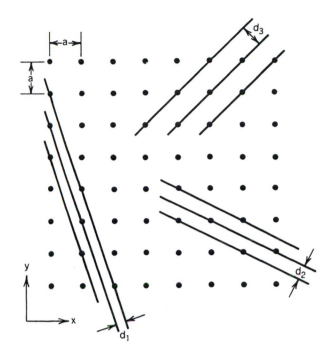

Figure 3

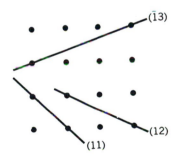

Figure 4

The Miller indices (hk) are used to specify a particular plane or a set of successive parallel planes. The (12) set of planes are shown in Figure 5 and it is not difficult to show

$$d = \frac{a}{\sqrt{5}} \qquad (5)$$

or

$$d = \frac{a}{\sqrt{1^2 + 2^2}} \qquad (6)$$

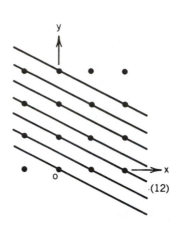

For a general set of planes in a square lattice with Miller indices (hk), the spacing d is

Figure 5

$$d = \frac{a}{\sqrt{h^2 + k^2}} \qquad (7)$$

For a three-dimensional, cubic crystal a general plane has intercepts qa, ra, and sa with the axes, and the Miller indices are (hkℓ) where h and k are given by Equation (4) and ℓ = m/s. As before, m is an integer and the three Miller indices are integers having no common factor > 1. The spacing d between successive planes of the (hkℓ) set is

$$d = \frac{a}{\sqrt{h^2 + k^2 + \ell^2}} \qquad (8)$$

The condition for constructive interference in terms of the lattice constant a and the Miller indices is obtained by combining Equations (3) and (8):

$$\frac{2a \sin \theta}{\sqrt{h^2 + k^2 + \ell^2}} = n\lambda \qquad (9)$$

The crystal structure of aluminum is face-centered cubic (Fcc), shown in Figure 6, having an aluminum ion at each corner and on each face of a cube. For such a structure, the condition for constructive interference is that h, k, and ℓ must be all even or all odd; other values of h, k, and ℓ do not satisfy Equation (9). Some of the Miller indices that give rise to constructive interference for a fcc structure are given in Table 1.

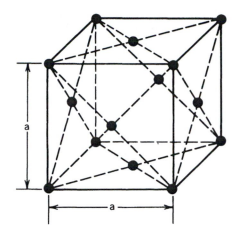

Figure 6

Table 1

h k ℓ	$h^2 + k^2 + \ell^2$	$(h^2 + k^2 + \ell^2)^{1/2}$
1 1 1	3	1.732
2 0 0	4	2.000
2 2 0	8	2.828
3 1 1	11	3.316
2 2 2	12	3.464
4 0 0	16	4.000
3 3 1	19	4.358

Question 2

In Experiment 6 you studied the particle properties of a car on an air track and in Experiment 13 you studied the particle properties of an electron in a CRT. Use your data from Experiment 6 to calculate the wavelength associated with the car. Use your answer to part a of Question 1, Experiment 13, to calculate the wavelength of the electron in the CRT. Briefly explain why it is or is not possible to experimentally determine each wavelength.

Outcomes

After you have finished the activities in this experiment you will have·

a. studied de Broglie's hypothesis.
b. established the wave nature of the electron.
c. varied the electron wavelength with accelerating voltage.
d. calculated the lattice spacing of one or more crystals.

Experiment

Electron diffraction tubes usually have a target of graphite or polycrystalline aluminum, and some tubes have both targets where the electron beam may be focused on either one. Experiments will be described for both targets.

Polycrystalline Aluminum

The target is composed of many small crystals of aluminum and the orientation of these crystals is nearly continuous. For some wavelength, determined by the accelerating voltage of the tube, each crystal having a particular set of (hkℓ) planes which happen to be oriented such that Bragg's law is satisfied will give rise to con-structive interference. Since the orientation of this set of planes is con-tinuous the interference pattern will be a circle. Figure 7 shows two ring dif-fraction patterns produced by the con-structive interference of electron waves diffracted by two different sets of planes. If we change λ, then the "old" ring pattern will disappear and one or more new ring patterns will appear due to waves diffracted from a different set of planes.

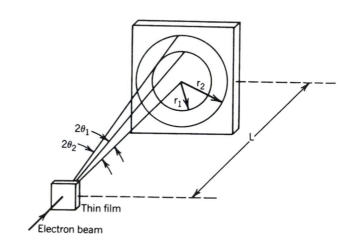

Figure 7

CAUTION. The accelerating voltage is typically a few thousand volts; therefore, do not touch any wires connected to the tube. Check with your instructor for operating instructions before turning on the equipment.

Measure the diameter of all observed ring diffraction patterns as a function of the accelerating voltage. From the observed ring diameter and the known value of L (target to screen distance, available from your instructor) calculate sin θ for each ring. It is suggested below that you first analyze your data and determine the lattice constant a.

For each voltage compute λ using Equation (2). For each value of θ use Bragg's law, Equation (9), with n = 1 to compute the lattice constant a for all possible reflecting planes in Table 1. The lattice constant a is then the consistent value that occurs for each value of θ. Calculate an average value from the consistent values.

Question 3

What is the percent discrepancy between your average value and the value determined from X-ray diffraction, 4.041 Å?

Another type of analysis is to assume the lattice constant is known from X-ray data, then calculate the electron wavelength using Bragg's law and compare it with the value computed from de Broglie's postulate. This is the sort of analysis that Davisson and Germer did in 1927. Analyze your data in this manner. As before, take n = 1 in Bragg's law and for each voltage compute λ for each plane given in Table 1. For each voltage a value of λ calculated from Bragg's law should be consistent with the value computed from de Broglie's equation.

Question 4

Do you find your experimental results are adequately explained by the wave properties of electrons? Briefly explain.

Graphite

The single graphite crystal produces a hexagonal spot diffraction pattern on the tube screen as shown in Figure 8. The distance r from the central spot (undiffracted beam) to each of the spots on the hexagon is measured. The separation d between planes is related to λ, L, and r by Bragg's law and geometry:

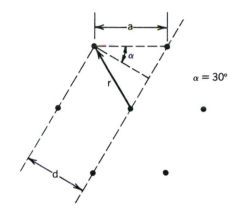

$$\lambda = 2d \sin\theta \cong 2d\theta \tag{10}$$

$$\frac{r}{L} = \tan 2\theta \cong 2\theta$$

where the small angle approximation was used and, as before L is the target-screen distance. Eliminating θ and solving for d:

$$d \cong \frac{\lambda L}{r} \tag{11}$$

Figure 8

From the geometry the lattice constant a may be related to λ, L, and r:

$$a = \frac{\lambda L}{r \cos \alpha} \tag{12}$$

If the small angle approximation is not valid for your apparatus, then you should derive an equation analogous to Equation (12) which does not make this approximation.

The lattice constant a may be determined by the following. Measure r for several values of the accelerating voltage V. For each voltage V calculate the de Broglie wavelength using Equation (2) and then knowing L calculate the lattice constant a using Equation (12) or the analogous equation which does not make the small angle approximation. Compute an average value for the lattice constant.

Question 5

What is the percent discrepancy between your average value and the value determined by X-ray diffraction, 2.456 Å?

As before, the data could be analyzed by assuming the lattice constant a is known from X-ray diffraction data and λ computed from Equation (12). Then the de Broglie wavelength could be computed using Equation (2) and the values of λ compared. Do such an analysis.

Question 6

Answer Question 4 for this experiment.

29
ATOMIC SPECTRA.
BALMER SERIES

Apparatus

Spectrometer, reflection grating (~30,000 lines/inch), converging lens with stand and lens holder, mercury vapor lamp, hydrogen discharge tube.

Introduction

A. HYDROGEN SPECTRA

Quantum theory applied to atoms predicts the energies of the atomic electrons are quantized, with a ground energy level (lowest energy level) and a constellation of excited energy levels. Electron transitions between these quantized energy levels results in the emission or absorption of photons having frequencies characteristic of each chemical element.

The energy levels for hydrogen may be obtained by using the Bohr theory. Often in beginning texts the mass of the nucleus is assumed infinitely large compared to the mass of the electron, so that the nucleus remains at rest. The more massive the nucleus the better the approximation, and even for hydrogen the approximation is good since the proton is 1836 times as massive as the electron. However, for very accurate spectroscopic measurements we must take into account the finite mass of the nucleus.

For a finite mass nucleus the proton and electron revolve around a common center of mass as shown in Figure 1. It is not difficult to show that for a finite mass nucleus the motion is as if the proton was fixed and the electron orbits with a <u>reduced mass</u> μ where

$$\frac{1}{\mu} = \frac{1}{M} + \frac{1}{m} \tag{1}$$

or

$$\mu = \frac{Mm}{M + m} \tag{2}$$

This is shown in Figure 2. The reduced mass μ for hydrogen is about 0.05% smaller than the electron mass m. You should show this.

Figure 1   Figure 2

Applying the Bohr Theory to the system shown in Figure 2 yields the energy levels E_n for hydrogen with a nucleus of finite mass:

$$E_n = \frac{(2\pi)^2 \mu e^4}{(4\pi\epsilon_0)^2 2h^2 n^2} ; \qquad n = 1, 2, 3, \ldots \tag{3}$$

where h is Planck's constant, n is the principal quantum number, e is the magnitude of the charge of the electron, and ϵ_0 is the permittivity of empty space.

The energy levels for hydrogen and the transitions producing emission spectral lines are shown in Figure 3. Each energy level is specified by its principal quantum number n, where n = 1 is the ground energy level and n = 2, 3, 4, ... are the excited states.

It is possible to excite gas atoms by applying a high voltage to a discharge tube containing the gas at low pressure. Free electrons accelerated between the cathode and anode of the discharge tube produce collisional excitation of the hydrogen atoms. See Figure 4. Spontaneous de-excitation of the gas atoms results in emission spectral lines. The change in energy of the atomic electron is equal to the energy of the emitted photon:

$$E_{n_i} - E_{n_f} = h\nu = \frac{hc}{\lambda}, \qquad n_i = n_f+1, \; n_f+2, \ldots \tag{4}$$

where ν and λ are the frequency and wavelength of the photon and c is the velocity of light. Solving Equation (4) for the reciprocal of λ:

$$\frac{1}{\lambda} = \frac{1}{hc} (E_{n_i} - E_{n_f}) \tag{5}$$

and using Equation (3):

$$\frac{1}{\lambda} = \frac{(2\pi)^2 \mu e^4}{(4\pi\varepsilon_0)^2 hc2h^2} \left(-\frac{1}{n_i^2} + \frac{1}{n_f^2}\right) \tag{6}$$

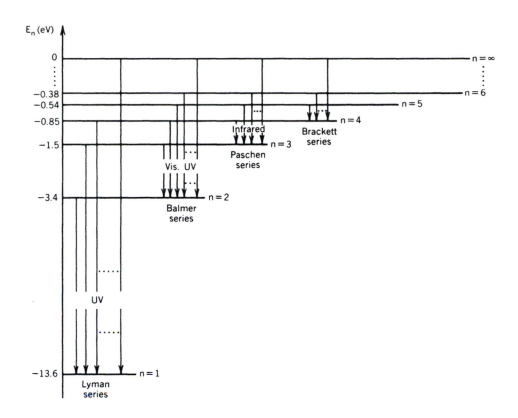

Figure 3

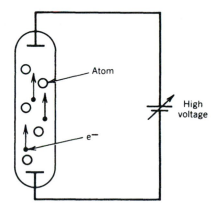

Figure 4

be: The Rydberg constant R_H for a hydrogen nucleus of finite mass M is defined to be:

$$R_H \equiv \frac{(2\pi)^2 \mu e^4}{(4\pi\varepsilon_0)^2 hc2h^2} \qquad (7)$$

Equation (6) may then be written

$$\frac{1}{\lambda} = R_H \left(\frac{1}{n_f^2} - \frac{1}{n_i^2} \right) \qquad (8)$$

Evaluating R_H using the currently accepted values of m, M, e, h, c, and ε_0, we find

$$R_H = 1.09681 \times 10^7 \ m^{-1} \qquad (9)$$

The Rydberg constant R_∞ for an assumed infinitely massive nucleus is given by:

$$R_\infty = \frac{m}{\mu} R_H = 1.09741 \times 10^7 \ m^{-1} \qquad (10)$$

The experimental value for the Rydberg constant R_H, determined by very accurate spectroscopic measurements is

$$R_H \pm \delta R_H = 10967757.6 \pm 1.2 \ m^{-1} \qquad (11)$$

Hence the Rydberg constant for a finite mass nucleus gives better agreement with the spectroscopic data than R_∞.

B.   REFLECTION GRATING AND SPECTROMETER

     Wavelength measurements are among the most accurate measurements in physics. The tool of measurement in this experiment is the spectrometer, principally a device for the accurate measurement of the angle of deflection of light (whether by reflection, refraction, or diffraction). The spectrometer was used to determine the index of refraction of a glass prism in Experiment 22. Figure 8 of Experiment 22 is a photograph of the spectrometer. Figure 5 shows the spectrometer viewed from above.

     The reflection grating is an aluminized glass surface with precisely cut, parallel grooves. The number of reflecting surfaces per inch (1/d) is approximately 30,000. See Figure 6.

     NEVER TOUCH OR RUB THE SURFACE OF THE GRATING, TO DO SO DESTROYS THE GRATING.

Source

Lens

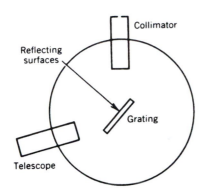

Collimator

Reflecting
surfaces

Grating

Telescope

Figure 5

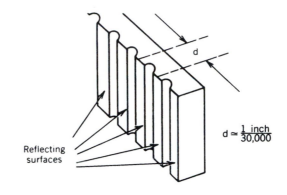

$d \simeq \frac{1 \text{ inch}}{30,000}$

Reflecting
surfaces

Figure 6

Each surface acts as a source of spherical waves (Huygen's principle) as shown in Figure 7. If the path difference from source to detector (the eye) of rays 1 and 2 is $m\lambda$, where $m = 0, 1, 2, \ldots$, then constructive interference occurs. Note in Figure 8 that ray 2 travels a distance x_2 farther than ray 1 to reach the grating, where

$$x_2 = d \cos \alpha = d \cos (\frac{\pi}{2} - \theta_{in}) = d \sin \theta_{in}$$

(12)

Whereas, ray 1 travels a distance x_1 farther than ray 2 from the grating to reach the detector, where

$$x_1 = d \cos \beta = d \cos (\frac{\pi}{2} - \theta_{out}) = d \sin \theta_{out}$$

(13)

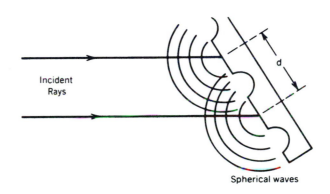

Incident
Rays

Spherical waves

Figure 7

The path difference is $x_2 - x_1$ and for constructive interference:

$$x_2 - x_1 = d(\sin \theta_{in} - \sin \theta_{out}) = m\lambda$$

(14)

Note that $\theta_{in} = \theta_{out}$ implies $m = 0$, which is the central maximum intensity.

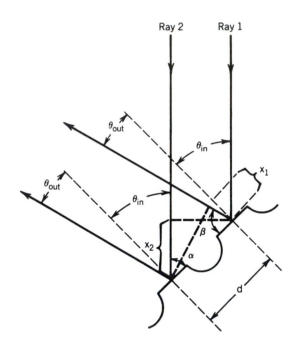

Figure 8

Outcomes

After you have finished the activities in this experiment you will have:

a. become familiar with the grating spectrometer.

b. calibrated your reflection grating by determining d.

c. studied the Balmer series of hydrogen.

Alignment of Spectrometer

First, move the mercury lamp near the slit and adjust it vertically. Move the lamp to about 20 cm from the slit and focus the light on the slit using the lens. The spectrometer will be aligned such that the angle θ_{in} is 55° (arbitrarily). The steps to do this are numbered below and in Figure 9.

1. With the grating removed focus the telescope crosshairs on the nonmovable edge of the collimator slit.

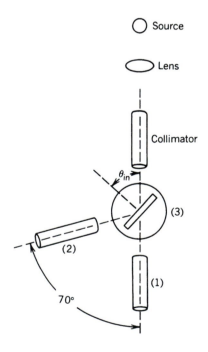

Figure 9

2. Then rotate the telescope 70° and read its angular position.

3. Then place the grating on the table and rotate the table until the non-movable edge of the slit is centered on the crosshairs. You are viewing the m = 0 spectrum in this position. The m = 1 spectral lines are to the left and m = -1 to the right.

Calibration of Grating

The grating you are using is a replica; therefore, it has approximately the number of lines per inch of the original grating. Hence we must use known wavelengths to calibrate the grating. It is suggested that the mercury arc lamp be used as the source of known wavelengths.

Qualitatively observe the mercury spectrum in first order for m = +1, identifying as many lines as possible. To observe fainter lines it may be necessary to increase the slit width. The color and wavelength of some spectral lines of mercury are given in Table 1 of Experiment 22.

Choose 5 or 6 lines; for each line measure the angular position for m = +1 and determine θ_{out}. Plot λ versus $\sin \theta_{out}$, where the values for λ are taken from Table 1 of Experiment 22. From your graph determine d and θ_{in}. If a computer is available, determine d and θ_{in} by a least squares fit.

Question 1

From your graph how could you determine if an incorrect value of λ was assigned to a line? Did you make such a mistake?

Balmer Series

Replace the mercury lamp with the discharge tube containing low pressure hydrogen gas. THE VOLTAGE SOURCE FOR THE DISCHARGE TUBE IS A FEW THOUSAND VOLTS. BE CAREFUL! Measure the first-order angular position of each line of the Balmer series in the visible part of the spectrum. Calculate each wavelength. Accepted values are given in Table 1.

Table 1

Balmer Series

Color	λ (nm)	Transition
red	656.28	3 → 2
blue-green	486.13	4 → 2
blue	434.05	5 → 2
violet	410.47	6 → 2

Question 2

What is the percent discrepancy between each experimental value and the accepted values given in the table?

Plot $\frac{1}{\lambda}$ against $(1/4 - 1/n^2)$.

Question 3

What is the value of R_H determined from your graph?

Compare your value of R_H with the accepted value, $1.0968 \times 10^7 \ m^{-1}$, and with R_∞, $1.0974 \times 10^7 \ m^{-1}$, by calculating the percent discrepancy in each case.

Question 4

Are your measurements accurate enough to detect the finite mass of the nucleus?

30
ATOMIC SPECTRA. HELIUM, NEON, HELIUM-NEON LASER

Apparatus

Spectrometer, reflection grating ($\sim$30,000 reflecting surfaces/inch), helium and neon discharge tubes, discharge tube power supply, helium-neon 0.5-mW laser, lab jack, prism with stand, mercury vapor lamp, masking tape.

Introduction

Some interactions of radiation with atomic electrons of matter are shown in Figure 1. For simplicity only two electronic energy states are shown: E_1 is the ground state and E_2 is the first excited state. In each interaction shown in Figure 1, the photon energy, hf, is equal to the change in electron energy:

$$hf = E_2 - E_1 \qquad (1)$$

Spontaneous emission is the transition of an electron from a higher energy state to a lower state with the emission of a photon. When many atoms, such as those in a discharge tube, undergo spontaneous emission the emitted photons have random directions and phases, and the radiation is said to be incoherent. In the

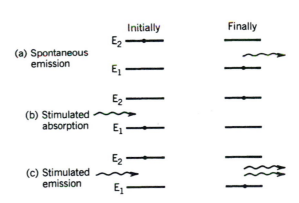

Figure 1

stimulated absorption process, an incident photon is absorbed and the electron makes a transition to an excited state. In the stimulated emission process, the incident photon stimulates the electron to make a transition to the ground state with the emergence of two identical photons, the original incident photon and the emitted photon. The two photons have the same phase, frequency, and direction, and a beam of such photons is said to be coherent radiation.

The mean lifetime of an electron in most excited states is about 10^{-8} s, i.e., after about 10^{-8} s the electron makes a transition from the excited state to a lower energy state. For some excited states, called metastable states, the decay is much slower and electron lifetimes in metastable states may be as long as 10^{-3} s.

Note that the photons emerging from stimulated emission may later be absorbed by other atoms, producing stimulated absorption. Thus stimulated absorption reduces the intensity of coherent radiation, whereas stimulated emission increases the intensity of coherent radiation. Quantum theory predicts that the probability per second per atom of stimulated emission occurring is equal to that of stimulated absorption. Hence if initially the number of atoms having an electron in state E_2 exceeds the number of atoms having an electron in state E_1, i.e., population inversion exists, then later the intensity of the beam of coherent radiation will have been amplified, i.e., an increase in the number of identical photons.

A laser, an acronym for "light amplification by stimulated emission of radiation," is a device for creating a population inversion and thus producing a beam of coherent radiation.

The basic components of a continuous laser are:

a. an "active material" which includes the atoms that interact with radiation.

b. a mechanism to create a greater population of electrons in an excited state than exist in a lower state.

c. a resonator or cavity that has a resonant frequency equal to the photon frequency.

In the atomic laser that operates with a mixture of helium and neon gases, the basic components are:

a. the active material is the gas composed of helium and neon atoms. Simplified, but adequate for this discussion, energy states for helium and neon are shown in Figure 2a. The neon atoms account for the laser action.

b. the mechanism for producing a greater population of electrons in the 5S states, as compared to the 3P states of neon is an electrical discharge from an applied DC voltage, followed by helium-neon atomic collisions.

c. the two mirrors separated by a distance ℓ form a resonant cavity. The separation of the two mirrors must be chosen so that constructive inter-ference occurs inside the cavity. The terms of waves, the two mirrors must produce standing waves inside the cavity.

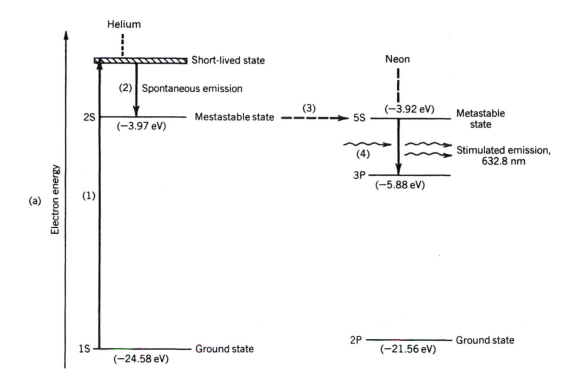

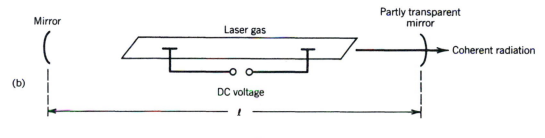

Figure 2

(a) Relevant energy states of He and Ne are shown and the major steps in the lasing process are:

(1) The electrical discharge primarily excites He atoms to short lived states.

(2) Spontaneous decay to metastable 2S states occurs.

(3) Collision of He and Ne atoms transfers the electron from the 2S state of He to the metastable 5S state of Ne.

(4) Stimulated emission occurs between the 5S and 3P states of Ne.

(5) Electrons are removed from the 3P state by collisions of Ne atoms with the walls which contain the gas. Steps 1 through 5 are continuously repeated when the laser is operating.

(b) The mirrors separated by a distance ℓ form the cavity. Standing waves exist between the mirrors and some fraction of the traveling waves incident on the partly transparent mirror is transmitted.

In this experiment you will use a spectrometer with a reflection grating to observe the emission spectral lines of helium and neon. After observing the neon spectrum, you will set up the helium-neon laser so that you can simultaneously observe the laser light through the spectrometer.

Outcomes

After finishing the activities in the experiment you will have:

a. measured some of the wavelengths of the spectral lines emitted by both helium and neon and compared your measured values with accepted values.

b. observed the wavelength emitted by a helium-neon laser and compared it with the wavelengths observed in the neon spectrum.

c. gained understanding of the basic mechanisms involved in the lasing process.

Alignment of Spectrometer and Calibration of Grating

Following the procedure of the previous experiment, align the spectrometer and then use the 576.96-nm, yellow line of mercury to calibrate the reflection grating.

Helium Spectra

Replace the mercury lamp with the discharge tube containing the low pressure helium gas. Spectral lines of helium are given in Table 1. THE VOLTAGE SOURCE FOR THE DISCHARGE TUBE IS A FEW THOUSAND VOLTS. BE CAREFUL!

Choose three lines and measure the first-order angular position of each line. Calculate the wavelength of each line and its error.

Question 1

Do the wavelengths determined experimentally agree with the accepted values given in Table 1?

Neon Spectra

Replace the helium discharge tube with the neon tube. Choose three lines, measure their angular positions in first order, and calculate the wavelengths. Spectral lines of neon are given in Table 2.

Question 2

Do the accepted and measured wavelengths agree within the experimental errors?

Table 1

Spectral Lines of Helium (Å)

Intensity	Wavelength	Intensity	Wavelength
1	4009.27	4	4713.38
50	4026.191	20	4921.931
5	4026.36	100	5015.678
12	4120.82	10	5047.74
2	4120.99	500	5875.62
3	4143.76	100	5875.97
10	4387.929	100	6678.15
3	4437.55	3	6867.48
200	4471.479	200	7065.19
25	4471.68	30	7065.71
30	4713.146		

Table 2

Spectral Lines of Neon (Å)

Intensity	Wavelength	Intensity	Wavelength
10	4537.754	100	5944.834
10	4540.380	100	5965.471
15	4704.395	100	5974.627
12	4708.862	120	5975.534
10	4710.067	80	5987.907
10	4712.066	100	6029.997
15	4715.347	100	6074.338
10	4752.732	80	6096.163
12	4788.927	60	6128.450
10	4790.22	100	6143.063
10	4827.344	120	6163.594
10	4884.917	250	6182.146
4	5005.159	150	6217.281
10	5037.751	150	6266.495
10	5144.938	60	6304.789
25	5330.778	100	6334.428
20	5341.094	120	6382.992
8	5343.283	200	6402.246
60	5400.562	150	6506.528
5	5562.766	60	6532.882
10	5656.659	150	6598.953
5	5719.225	70	6652.093
12	5748.298	90	6678.276
80	5764.419	20	6717.043
12	5804.450	100	6929.467
40	5820.156	90	7024.050
500	5852.488	100	7032.413
100	5872.828	50	7051.292
100	5881.895	80	7059.107
60	5902.462	100	7173.938
60	5906.429		

Helium-Neon Laser

CAUTION: The helium-neon laser line should be observed with great caution to avoid any
possiblity of eye damage.

Use a prism to reflect the laser
as shown in Figure 3. Do NOT replace
the prism with a mirror. Attach tape to
the other two sides of the prism to reduce
the scattering of laser light throughout
the laboratory.

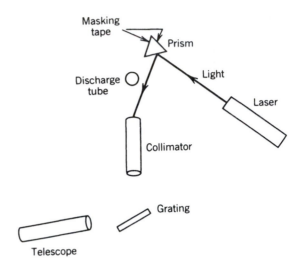

Leave the neon tube so that its
spectrum can be observed through the
spectrometer. Calculate the angular
position of the first order, 632.8 nm
helium-neon laser line. Rotate the
telescope to this angular position.
CLOSE THE COLLIMATOR SLIT. Arrange
the helium-neon laser so that the laser
and neon light fall on the CLOSED SLIT.
This can be accomplished as shown in
Figure 3, where some of the laser light
is reflected from the prism to the
CLOSED SLIT. OPEN THE SLIT JUST
ENOUGH to observe the laser and neon
spectra.

Figure 3

Question 3

Does the laser line coincide with any of the neon lines? Explain why
it should or should not coincide. (Hint - During one lifetime of a metastable
state of neon, how many short-lived states of neon decayed?)

Question 4

The lifetimes of a short-lived state, a metastable state, and the ground
state are typically 10^{-8} s, 10^{-3} s, and effectively infinite. Using the
Heisenberg energy-time, uncertainty principle, what is the uncertainty ΔE
in the energy of each state? This energy spread of state is usually called
the <u>width</u> of the state.

Question 5

Assuming the 3P state of neon (see Figure 2a) has a sufficiently long
lifetime that its width is negligible compared to the width of the 5S state,
calculate the uncertainty Δf in the frequency of the emitted photon.

This uncertainty is the so-called <u>natural width</u> of the spectral line. Usually other
mechanisms such as Doppler broadening and pressure broadening due to the collisions of
atoms in the source, cause the line to be much broader than its natural width.

Question 6

Is the natural width of the 632.8-nm line significant? To answer this
question calculate the fractional width of the line, $\Delta f/f$.

31
RADIOACTIVE DECAY. STATISTICS, HALF-LIFE

Apparatus

Geiger-Mueller tube and sample holder, scaler/timer, oscilloscope, cesium-137 sample. For the entire lab: indium-113 sample, Geiger-Mueller tube and auxiliary equipment.

Introduction

A. SPONTANEOUS DECAY OF RADIONUCLIDES

An unstable nucleus typically decays in one of six ways in order to achieve a new configuration that is either stable itself or will lead to one that is stable. For each mode of decay the nuclear reactions are summarized below, where $_Z^A X$ specifies the "parent" chemical element X with mass number A and atomic number Z.

Alpha decay is the emission of a helium-four nucleus (alpha particle)

$$_Z^A X \longrightarrow \, _{Z-2}^{A-4} Y + \, _2^4 He \tag{1}$$

where Y specifies the "daughter" chemical element. The kinetic energies of the alpha particles emitted by a given source are discrete and typically 4 to 7 MeV.

Beta decay is the emission of an electron (β^-) and an anti-neutrino ($\bar{\nu}_e$) or a positron (β^+) and a neutrino (ν_e):

$$\underset{Z}{\overset{A}{}}X \longrightarrow \underset{Z+1}{\overset{A}{}}Y + \beta^- + \bar{\nu}_e \qquad (2)$$

or

$$\underset{Z}{\overset{A}{}}X \longrightarrow \underset{Z-1}{\overset{A}{}}Y + \beta^+ + \nu_e \qquad (3)$$

A beta source emits beta particles having kinetic energies ranging from zero up to a definite maximum. The energy not given to a beta particle is carried away by the neutrino or anti-neutrino. The maximum kinetic energy is typically somewhat less than 1 MeV, but may be more than 10 MeV.

Gamma decay is the emission of a high frequency photon:

$$\underset{Z}{\overset{A}{}}X^* \longrightarrow \underset{Z}{\overset{A}{}}X + \gamma \qquad (4)$$

where the asterisk indicates the parent nucleus is in an excited state. A gamma source emits gamma rays having discrete energies up to about 10 MeV. Most decays of other modes leave the nucleus in an excited state, so gamma rays usually accompany any other radiation.

Electron capture is the capture and annihilation of an orbital electron by the nucleus and the emission of a neutrino:

$$\underset{Z}{\overset{A}{}}X + e^- \longrightarrow \underset{Z-1}{\overset{A}{}}Y + \nu_e \qquad (5)$$

It is typically followed by X-ray emission as the daughter atom moves to the ground state.

Internal conversion is the transfer of excess nuclear energy to an atomic electron causing the electron to be ejected from the atom. The nucleus decays to a lower energy state.

$$\underset{Z}{\overset{A}{}}X^* + e^- \longrightarrow \underset{Z}{\overset{A}{}}X + e^- \qquad (6)$$

Fission is the splitting apart of a heavy nucleus into two nearly equal fragments:

$$\underset{Z}{\overset{A}{}}X \longrightarrow \underset{Z_1}{\overset{A_1}{}}Y_1 + \underset{Z_2}{\overset{A_2}{}}Y_2 + f\underset{1}{\overset{0}{}}n \qquad (7)$$

where $\underset{1}{\overset{0}{}}n$ is the symbol for the neutron and f is the number of neutrons produced, typically 1 to 3. Spontaneous fission is rare among the naturally occurring elements.

In this experiment you will study one sample which undergoes beta decay, and another sample which undergoes gamma decay. The suggested source of beta particles is cesium-137 which decays to barium-137:

$$\underset{55}{\overset{137}{}}Cs \longrightarrow \underset{56}{\overset{137}{}}Ba^* + \beta^- + \bar{\nu}_e \qquad (8)$$

Following beta or alpha decay the resulting nucleus is often in an excited state and it may decay by gamma ray emission. In this case

$$^{137}_{56}Ba* \longrightarrow \,^{137}_{56}Ba + \gamma \qquad (9)$$

The decay of cesium-137 to barium-137 is shown in terms of energy levels in Figure 1. The specified energy of each beta ray is the maximum value.

The suggested gamma ray source is discussed in Part C.

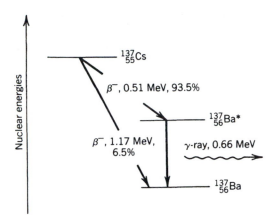

Figure 1

B. RANDOMNESS IN RADIOACTIVE DECAY: POISSON DISTRIBUTION

The decay of a radioactive nucleus is a random event and we cannot predict when any particular nucleus will decay. If we have a sample containing a very large number of radioactive nuclei, then the idealized probability that n will decay during a specified time interval is given by the Poisson distribution

$$P(n) = \frac{(\bar{n})^n \, e^{-(\bar{n})}}{n!} \qquad (10)$$

where $\bar{n}$ is the average number of decays in the interval. For small $\bar{n}$, P(n) versus n is a lopsided distribution. As $\bar{n}$ increases the Poisson distribution approaches the symmetrical Gauss distribution (see GAUSS DISTRIBUTION in the INTRODUCTION). In Figure 2 a Poisson distribution is sketched for two values of $\bar{n}$.

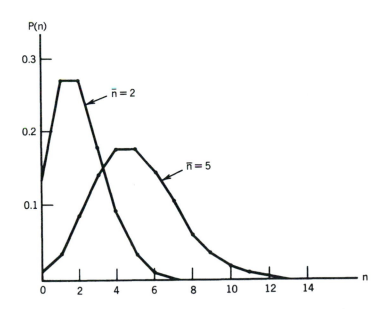

Figure 2

Suppose we have a sample of radioactive nuclei and we carry out a large number (M) of identical, independent measurements of the count rate n (counts/unit time). The average count rate $\bar{n}$ is defined to be

$$\bar{n} = \frac{1}{M} \sum_{i=1}^{M} n_i \tag{11}$$

The standard deviation σ is defined as

$$\sigma = \left[\frac{1}{M-1} \sum_{i=1}^{M} (n_i - \bar{n})^2\right]^{\frac{1}{2}} \tag{12}$$

For a Poisson distribution, it can be shown that

$$\sigma = \sqrt{\bar{n}} \tag{13}$$

Assuming $\bar{n}$ is 100 counts/min then σ is 10. Approximately 68% of the observed count rates fall in the range

$$\bar{n} - \sigma \quad \text{to} \quad \bar{n} + \sigma \tag{14}$$

and approximately 95% will fall in the range

$$\bar{n} - 2\sigma \quad \text{to} \quad \bar{n} + 2\sigma \tag{15}$$

An idealized histogram is plotted in Figure 3. The distribution is a Poisson distribution. Notice that it resembles a Gauss distribution.

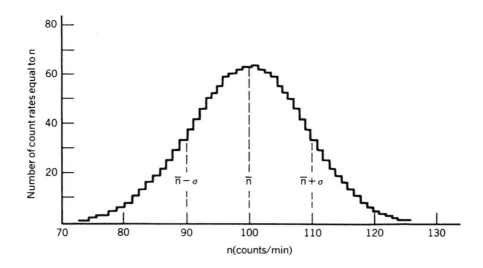

Figure 3

For any single measurement n, 68% of the time n will be within ±σ of the average count rate $\bar{n}$. We assign an approximate uncertainty to each single measurement n. If the single measurement is n (counts per unit time), the approximate uncertainty in n is ±$\sqrt{n}$. For each measurement the result to be reported is n ± $\sqrt{n}$.

C. HALF-LIFE

Every radionuclide has a characteristic half-life, the time required for half of the nuclei of a sample to decay. Some have half-lives of a millionth of a second; others have half-lives that range up to billions of years, e.g., the half-life of rubidium-87 is 4.8×10^{10} years. The half-life of ^{137}Cs is 30.0 years and the half-life of ^{137}Ba* is 2.55 minutes.

The decay of all radionuclides is exponential:

$$N(t) = N_0 e^{-0.693 \, t/T} \tag{16}$$

where N_0 is the number of parent nuclei present at t = 0, and N(t) is the number at time t, and T is the half-life. If we take the logarithm of both sides, we have

$$\log_{10} N = \log_{10} N_0 - \frac{0.693}{T} t \log_{10} e \tag{17}$$

Hence, a plot of $\log_{10} N$ vs. t would yield a straight line.

The suggested radionuclide to be studied is indium-113. Indium-113 is the "daughter" of the radionuclide tin-113. Tin-113 decays to an excited state of indium-113 by electron capture. Symbolically

$$^{113}_{50}Sn + e^- \longrightarrow {}^{113}_{49}In* + \nu_e \tag{18}$$

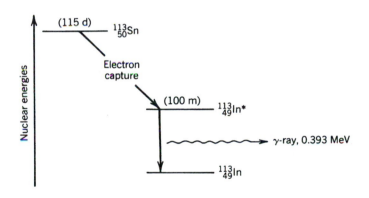

Figure 4

where ν_e is the neutrino and the asterisk implies indium-113 is in an excited state. It decays to the ground state by gamma ray emission:

$$^{113}_{49}In* \longrightarrow {}^{113}_{49}In + \gamma \tag{19}$$

Figure 4 shows the nuclear energy levels. The accepted half-life of ^{113}Sn is 115 days and that of ^{113}In* is 100 minutes.

D. DETECTION OF PARTICLES: GEIGER-MUELLER TUBE

Particles (α, β, and γ rays) are detected by interacting with matter. Charged particles are detected by the ionization produced and gamma rays are detected by conversion into electrons.

The Geiger-Mueller tube is a metallic cylinder with a metal wire along its axis and it has a "thin" window at one end to let in particles. The metallic cylinder and wire connect to a high voltage as shown in Figure 5.

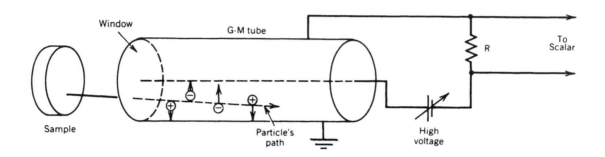

Figure 5

The tube is filled with a suitable gas, such as argon. Note the electrical circuit is completed through the gas. A charged particle, emitted by the radioactive sample, traversing the space between the wire and metal shell will ionize the gas atoms along its path. A neutral particle (photon, neutron) will occasionally interact to produce an electron and ion. The ions and electrons so produced will be accelerated toward the wire and shell. This slight ionization in the gas produced by the entrance of a single ionizing particle will produce secondary ionizations in that each ion and electron will produce ionizations as they accelerate to the electrodes (wire and shell). Hence a current pulse will occur in the circuit producing a voltage pulse across R. The resulting voltage pulse is amplified and registered in the scaler; thus individual particles entering the tube are counted. This process, from entrance of a single particle until it is counted, requires about 2×10^{-4} seconds, which is called the "dead time" of the tube since a second particle entering the tube during this time will not produce a separate pulse. A count every 2×10^{-4} seconds means a count rate of $1/2 \times 10^{-4}$ = 5000 counts/sec. Always keep your count rate below 50 counts/sec (3000 counts/min) so that you will not have to correct for dead time.

Beta rays easily penetrate the G-M tube window and are then detected. Alpha rays being much more massive and slower (for a given energy) and having twice the charge of beta rays lose energy much more rapidly in passing through matter (air, "window," etc.). Thus alpha rays have difficulty in penetrating the G-M tube window. Being electrically neutral, gamma rays penetrate matter more easily than alpha or beta rays. Because gamma rays may travel "large" distances without interacting with matter, the efficiency of G-M tubes for detecting gamma rays is only a few percent.

A photograph of the apparatus is shown in Figure 6.

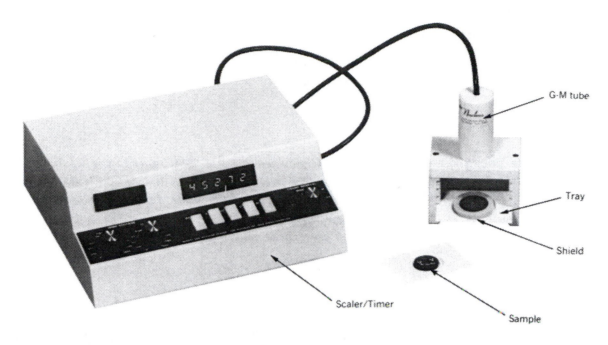

Figure 6

E. RADIOACTIVE SOURCE ACTIVITY

The activity of a radioactive source is specified in units of curies (Ci) where

$$1 \text{ curie} = 3.666 \times 10^{10} \text{ decays/sec} \qquad (20)$$

A G-M tube detects a fraction of the total activity of a radioactive source. Figure 7 shows a detector window of area A located a distance r from a small source. The number of particles per second n entering the detector is

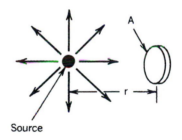

Figure 7

$$n = KS \frac{A}{4\pi r^2} \times 3.666 \times 10^{10} \qquad (21)$$

where K is the average number of particles emitted per decay and S is the source activity in curies.

Outcomes

After you have finished the activities in this experiment you will have:

a. studied the statistical nature of radioactive decay.
b. measured the half-life of a radioactive sample.

Operating Voltage of the Geiger-Mueller Tube

To establish the operating voltage place the suggested cesium sample in the tray beneath the tube as shown in Figure 6. Increase the high voltage until the scaler begins to count particles entering the tube. Then increase the voltage by 100 V. This is the operating voltage throughout the experiment.

Background Radiation

The background count arises from cosmic rays and radioactive building materials such as concrete. Place the cesium source well away from the Geiger tube and take a two-minute reading of background radiation. Express your result in counts per minute.

In most of the work to follow, you should subtract background count rate to obtain the significant count rate for a particular experiment.

Half-Life

Equation (16) gives the number of parent nuclei $N(t)$ remaining at time t, and $N(t)$ is a decreasing exponential. We measure the count rate n and we ask "How does $n(t)$ change with time t?" To answer this question we start by calculating the time derivative of Equation (16):

$$\frac{dN}{dt} = - \frac{0.693}{T} N_0 e^{-0.693\, t/T} = - \frac{0.693}{T} N(t) \tag{22}$$

This says the decay rate at time t is proportional to the number of nuclei at time t. Also the measured count rate n is proportional to the rate particles are emitted, and the rate of particle emission is proportional to the decay rate dN/dt. Thus the count rate $n(t)$ is proportional to $N(t)$. Hence we may write:

$$n(t) = n_0 e^{-0.693\, t/T} \tag{23}$$

It is suggested that one set of apparatus be used to take data for the entire lab. Place the indium-113 sample in the holder so that the count rate is less than 3000 counts/ minute. For about two hours take a one-minute count about every ten minutes. Students should take turns taking the data and recording it on the blackboard. Each student should record the data in his or her notebook. Subtract the background rate from each count rate. As you are accumulating data, commence plotting your corrected count rate vs. time on both linear and semilog graph paper.

Question 1

From your semilog graph, what is your value for the half-life? If a computer is available to do a least squares fit, then what is the computer calculated value for the half-life? Compare the two values.

Question 2

What is the percent discrepancy between your value and the accepted value? The accepted value is 100 minutes.

Question 3

If you had not corrected the count rate by subtracting the background rate, then would your value for the half-life increase, decrease, or not change?

Geiger-Mueller Tube Dead Time

The G-M tube is disabled for approximately 200 microseconds after it is discharged by the passage of a particle through it. If particles enter the tube at an interval less than this, then the measured counting rate will be too low.

To measure the dead time place the cesium-137 source near the G-M tube window to establish as high a counting rate as possible. Connect the detector output (located on the rear of the scaler) to the vertical input of the oscilloscope. Adjust the scope trigger and sweep time until the displayed voltage pulse (created by a particle entering the tube) resembles that shown in Figure 8. The dead time τ is shown in Figure 8. Measure the dead time for your tube.

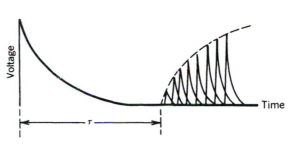

Figure 8

Question 4

If you want the uncertainty in the counts/sec, due to dead time, to be less than 1%, then what number of counts/sec should you not exceed?

Statistical Nature of Radioactive Decay

Place the cesium-137 sample so that about 10 counts/sec are detected and record the counts for 80 ten-second intervals. If time permits take more than 80 ten-second intervals. Plot your data in the form of a histogram, i.e., the number of ten-second intervals showing n counts vs. n. Use Equations (11) and (13) to calculate n̄ and σ. Calculate values of P(n) for the Poisson distribution using Equation (10). Multiply each value of P(n) by the total number of ten-second intervals and plot the points on the same graph as the histogram. Draw a smooth curve through the points. You do not need to calculate P(n) for each n; about 10 values of P(n) is adequate.

You do not need to subtract the background count rate, since it obeys the same statistical laws as the count rate n.

Question 5

What percent of your measurements fall in the range $\bar{n} \pm \sigma$?

Goodness of Fit (Optional)

How do we conclude whether the experimental data is fit by the Poisson distribution? A test for comparing the experimental and theoretical distributions is called "x^2 (chi-square) test for goodness of fit."

x^2 is defined by

$$x^2 = \sum_n \frac{[MP(n) - F(n)]^2}{MP(n)} \tag{24}$$

where M is the number of ten-second intervals, n is the number of counts in a ten-second interval, P(n) is given by Equation (10), and F(n) is the number of ten-second intervals showing n counts. Note for a given value of n the numerator in Equation (24) is the square of the frequency difference. The smaller x^2 is then the better the fit. Once x^2 is calculated standard tables are used to interpret the result. See, e.g., *Statistical Treatment of Experimental Data* by Hugh D. Young.

32

BETA AND GAMMA RAY ABSORPTION

Apparatus

Geiger-Mueller tube and sample holder, scaler/timer, lead absorbers, aluminum absorbers, cesium-137 sample, thallium-204 sample.

Introduction

A. GAMMA ABSORPTION

Gamma rays interact primarily with the electrons of matter. However, a gamma ray traversing matter does not gradually lose energy along its path but rather it interacts strongly at one point. The three principal ways gamma rays lose energy are:

Photoelectric absorption:	γ + atom $\longrightarrow$ ion + ejected electron
Compton scattering:	γ + atom $\longrightarrow$ γ' + ion + ejected electron
Pair production:	γ + atom $\longrightarrow$ atom + e^+ + e^-

All of the gamma ray energy is transferred to the ejected electron in photoelectric absorption. In Compton scattering some fraction of the incident gamma ray energy is transferred to the ejected electron. An electron-position pair is produced in pair production.

At low gamma ray energies ($\sim <$ 1 MeV) photoelectric absorption dominates. At intermediate energies ($\sim$ few MeV) compton scattering dominates, and pair production is the dominate mechanism for high energy gamma rays. Pair production will not occur if the gamma ray energy is less than 1.02 MeV. Why so?

Figure 1 shows a gamma ray beam of intensity I_0 incident on material of thickness dx and an emerging beam of intensity I. The fraction of the beam absorbed is proportional to dx:

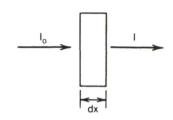

$$\frac{dI}{I} = -\mu dx \qquad (1)$$

where μ is the total absorption coefficient of the material and $dI = I - I_0 < 0$. Integrating Equation (1):

$$I(x) = I_0 e^{-\mu x} \qquad (2)$$

Figure 1

where $I(x)$ is the beam intensity after passing through material of thickness x. The total absorption coefficient is a sum of three terms:

$$\mu = \mu_{photo} + \mu_{Compton} + \mu_{pair} \qquad (3)$$

where μ_{photo} is the absorption coefficient due to photoelectric absorption, etc.

In general high density materials have a high density of electrons, hence they are better absorbers of gamma rays than low density materials. In order to compare the absorbing properties of different materials we define the "mass absorption coefficient" μ_m as

$$\mu_m \equiv \frac{\mu}{\rho} \qquad (4)$$

where ρ is the material density and we define the "mass thickness" x_m as

$$x_m = x\rho \qquad (5)$$

where x is the material thickness. We may then write the exponent in Equation (2) as

$$\mu x = \mu_m \rho x = \mu_m x_m \qquad (6)$$

where the units are: $\mu(cm^{-1})$, $x(cm)$, $\mu_m(cm^2/gm)$, and $x_m(gm/cm^2)$. The mass thickness of aluminum is

$$x_m = x \cdot 2.70 \qquad (7)$$

and the mass thickness of lead is

$$x_m = x \cdot 11.34 \qquad (8)$$

where the density of aluminum (2.70 g/cc) and lead (11.34 g/cc) were used. Thus for the same thickness x, lead has a mass thickness more than 4 times that of aluminum.

B. BETA ABSORPTION

Beta rays interact primarily with the electrons of matter via the coulomb interaction leaving a trail of ions and excited atoms. Unlike a gamma ray, a beta ray traversing matter usually loses energy gradually along its path. A difficulty in interpreting beta ray absorption data occurs because a beta source does not emit a monoenergetic beam, but rather the emitted betas have a continuous energy spectrum ranging from zero up to a definite maximum energy. It was pointed out in Experiment 31 that the neutrino or anti-neutrino carries away the difference between the maximum value and the beta ray energy.

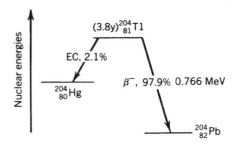

The suggested beta source in this experiment is thallium-204. The decay of thallium-204 is shown in Figure 2a, where the maximum energy of the emitted beta is specified. Figure 2b shows the continuous energy spectrum of the emitted betas.

Unlike gamma rays, the intensity of a beam of beta rays does not decrease exponentially with thickness of absorber.

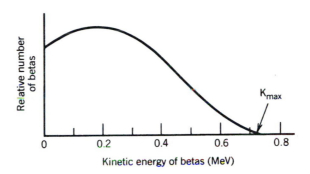

Figure 2

A useful relation between beta ray energy and range (maximum distance of penetration) is the empirically determined range-energy curve shown in Figure 3. Beta ray kinetic energy is plotted versus range in aluminum where the range is in mass thickness, mg/cm^2. For example, a range of approximately 13 mg/cm^2 stops a 0.1-MeV beta ray.

Outcomes

After finishing the activities in this experiment you will have:

a. measured gamma ray count rate as a function of lead absorber thickness.

b. determined mass absorption coefficient of lead for 0.66 MeV gamma rays.

c. measured beta ray count rate as a function of aluminum absorber thickness.

d. used a range-energy curve to determine the maximum energy of beta rays emitted by thallium-204.

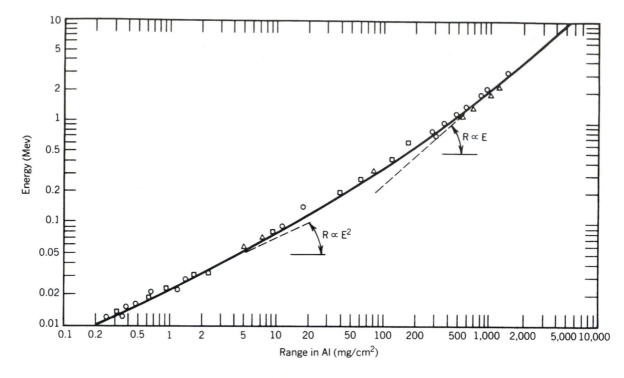

Figure 3

Absorption of Gamma Rays by Lead

Following the procedure in Experiment 31 use the cesium-137 sample to establish the operating voltage of a G-M tube and then measure the background count rate. Express your result in counts per minute.

The suggested gamma ray source is the cesium-137 sample. The decay of cesium-137 was shown in Figure 1 of Experiment 31. Note that it emits beta and gamma rays.

Place the cesium sample in the tray such that all lead absorbers may be stacked above it as shown in Figure 4. Leave the sample. Take about 8 one-minute counts as a function of absorber thickness, varying the absorber thickness from the thinnest to all absorbers stacked on each other. Subtract the background counts per minute from each data point.

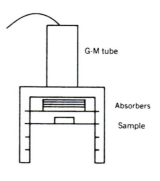

Figure 4

From Equation (2), the count rate n(x) as a function of absorber thickness x may be written

$$n(x) = n_0 e^{-\mu x} \qquad (9)$$

or in terms of mass thickness x_m

$$n(x_m) = n_0 e^{-\mu_m x_m} \qquad (10)$$

Absorbers are usually specified in mass thickness. Plot $n(x_m)$ versus x_m on appropriate graph paper such that a straight line is expected.

Question 1

Does your graph suggest the thinnest lead absorber stops the beta rays? Explain.

Question 2

What thickness x of lead is required to reduce the intensity of 0.66-MeV gamma rays by half?

Question 3

From your graph what is the value of the mass absorption coefficient of lead for 0.66-MeV gamma rays?

In Figure 5, photon mass absorption coefficient versus energy (100 keV to 1 MeV) is graphed for several elements, including lead. Knowing the mass absorption coefficient of lead, use Figure 5 to determine the gamma ray energy.

Question 4

What is the percent discrepancy between the accepted gamma ray energy (0.66 MeV) and your experimental value?

Absorption of Beta Rays by Aluminum

Place the thallium sample in the tray such that all aluminum absorbers may be stacked above it. Record one-minute counts as a function of absorber thickness. Correct each data point by subtracting background counts/min. Plot the corrected count rate versus mass thickness on semilog paper. You should not expect a straight line.

Question 5

What count rate would you expect to measure if the aluminum absorber thickness was such that the most energetic betas are barely absorbed?

Stack the aluminum absorbers until the count rate in Question 5 is obtained. Then, knowing the mass thickness required to stop the most energetic betas, use the range-energy curve in Figure 3 to determine the maximum energy of the betas emitted by thallium-204.

Question 6

What is the percent discrepancy between the accepted maximum beta ray energy, 0.766 MeV, and your experimental value?

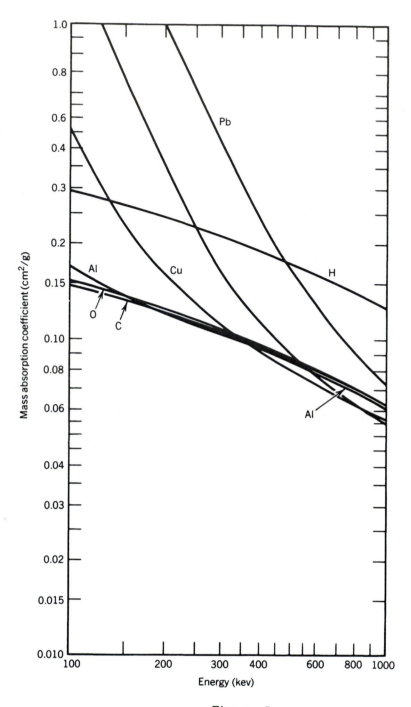

Figure 5

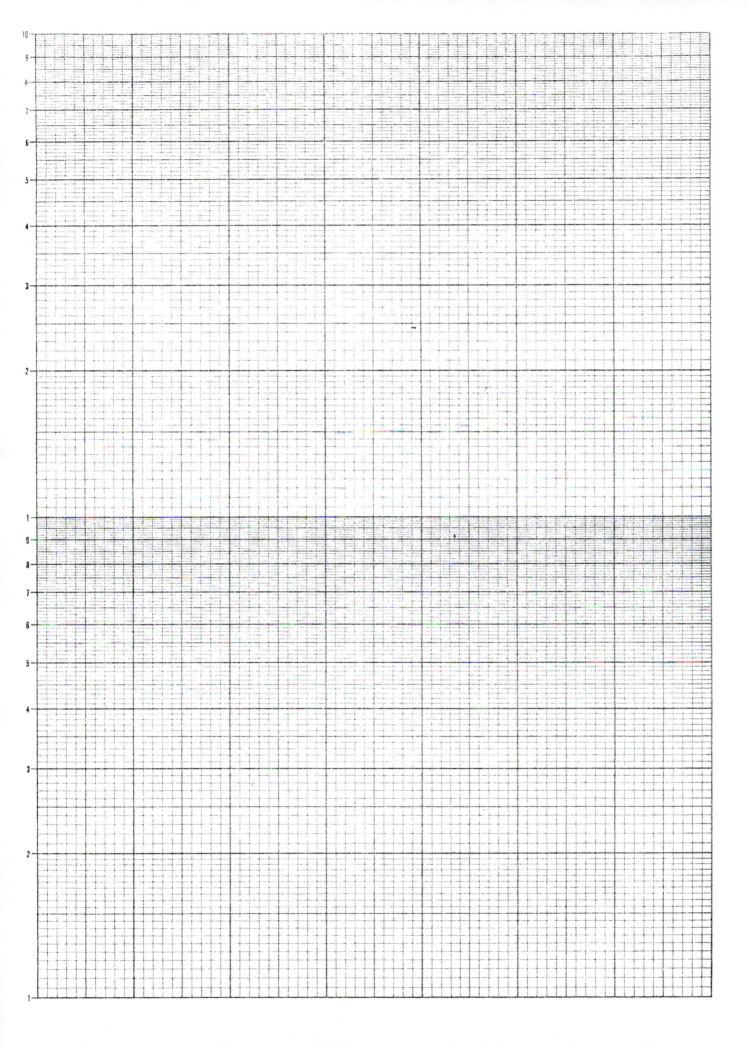

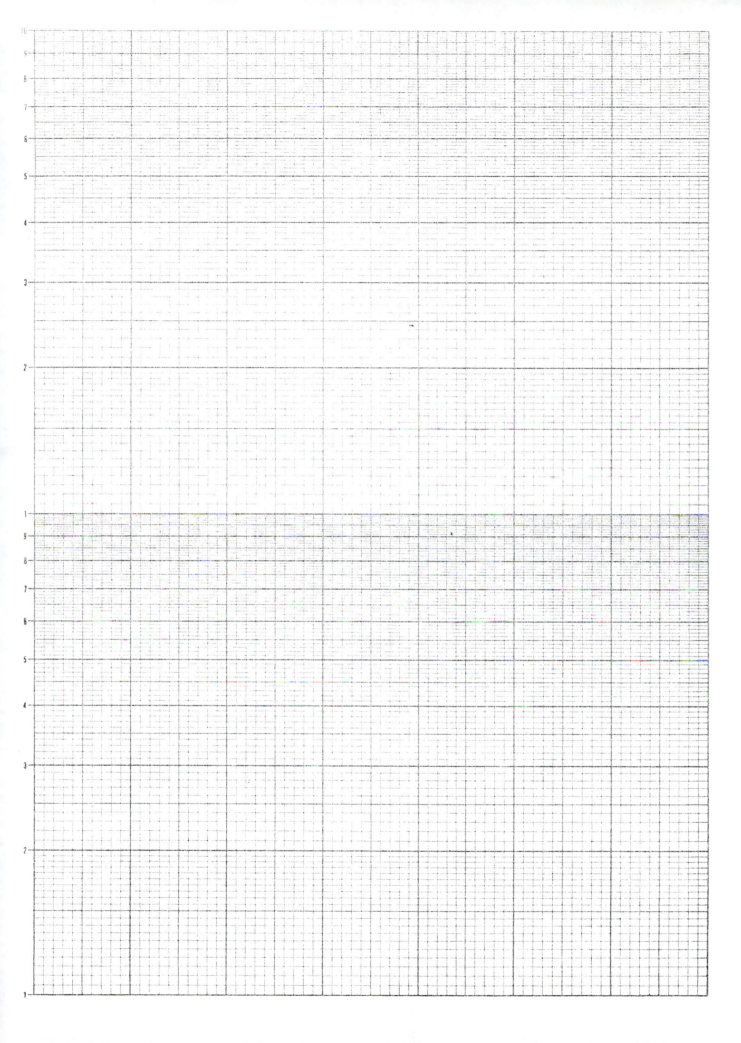

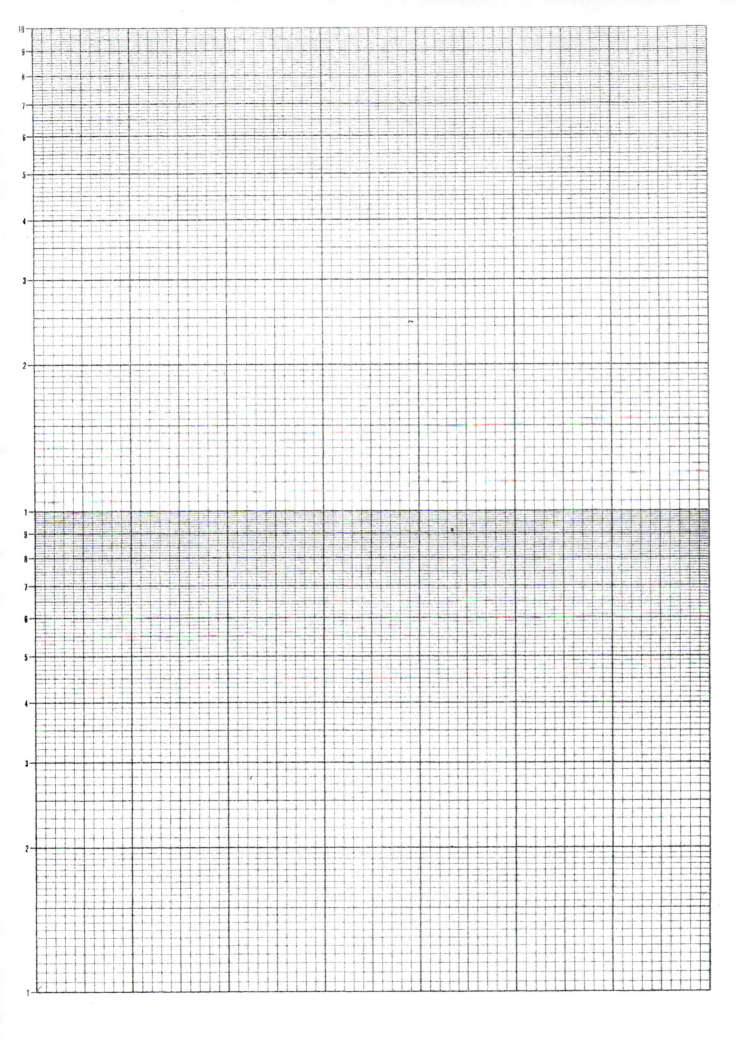

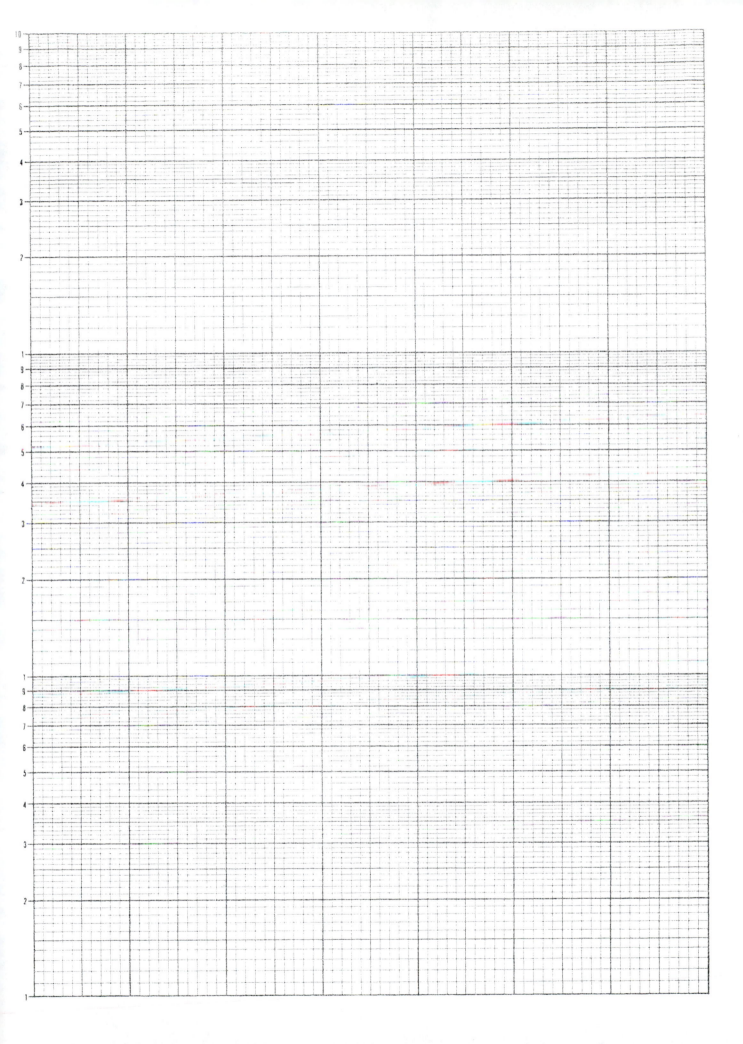

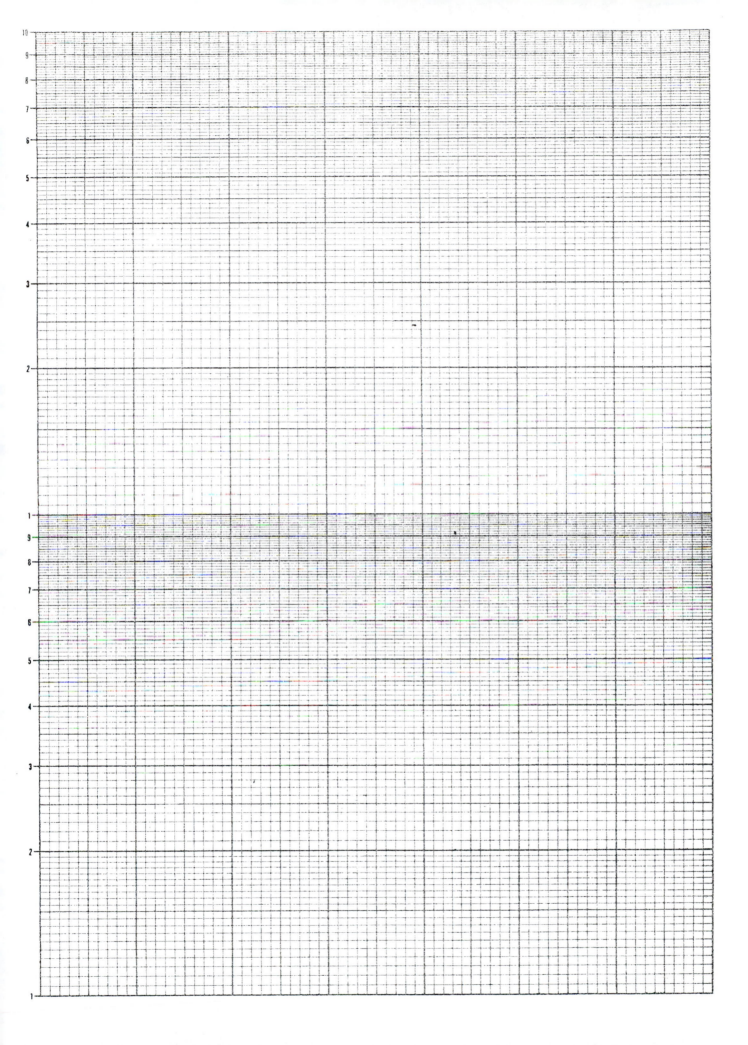

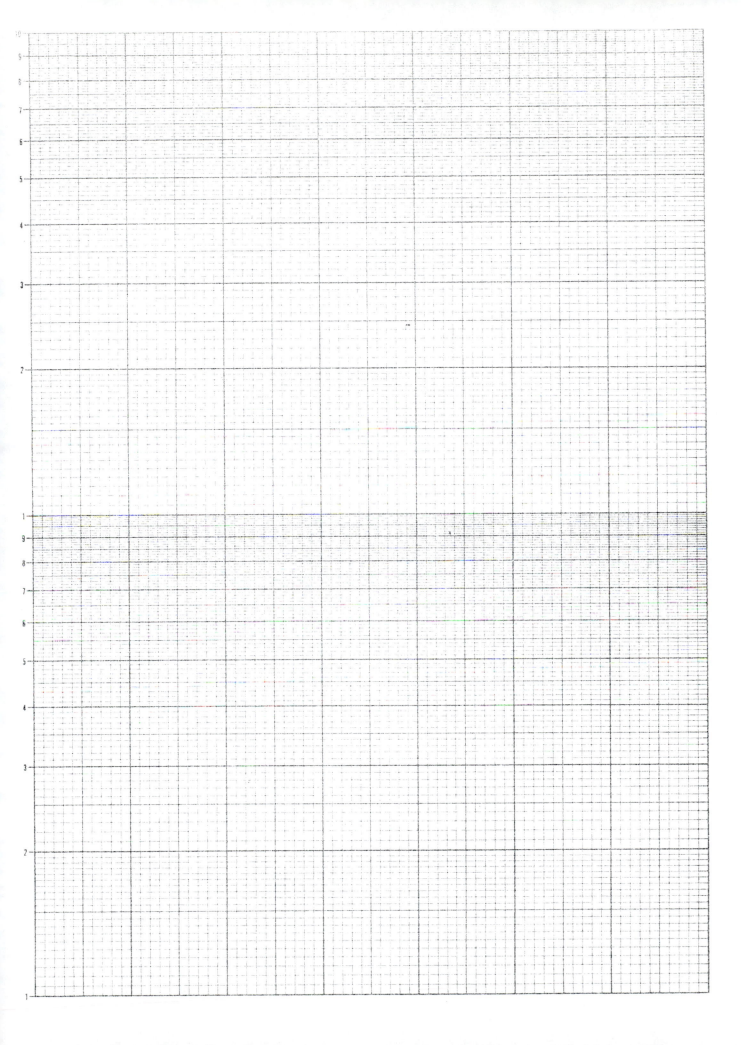

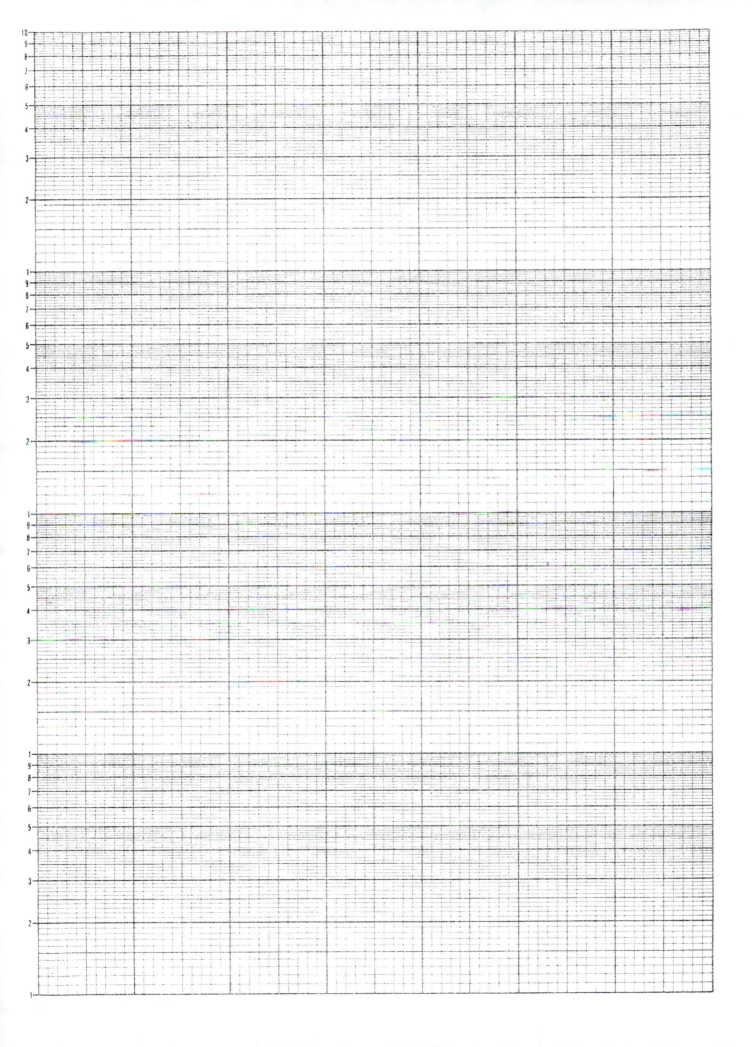

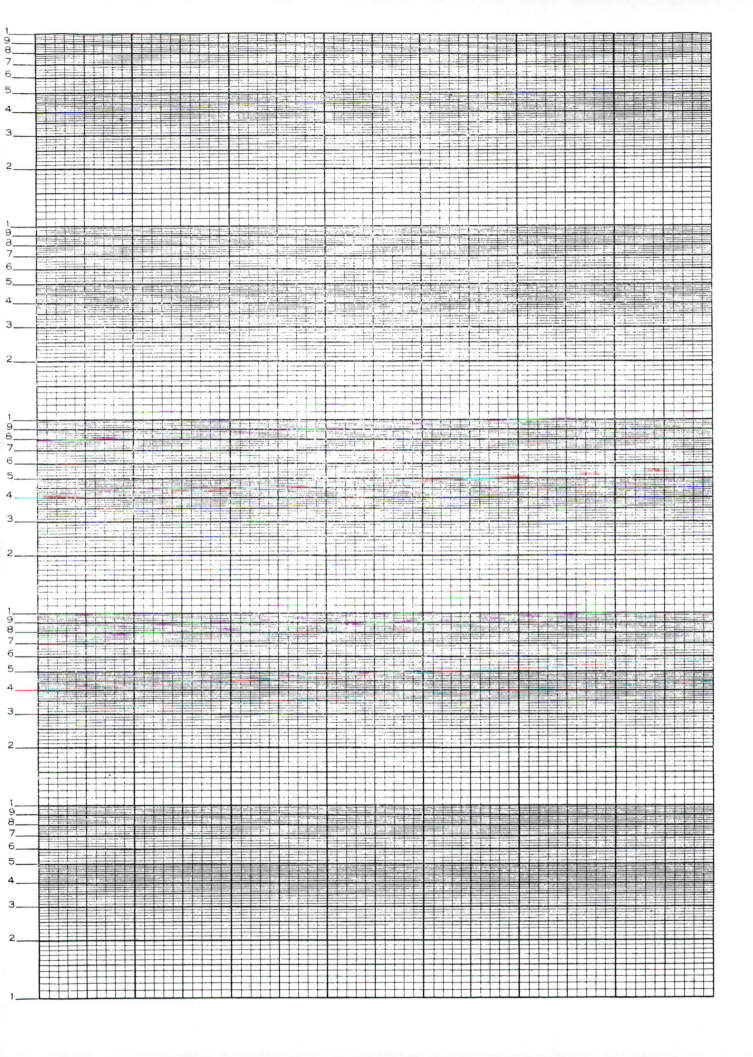

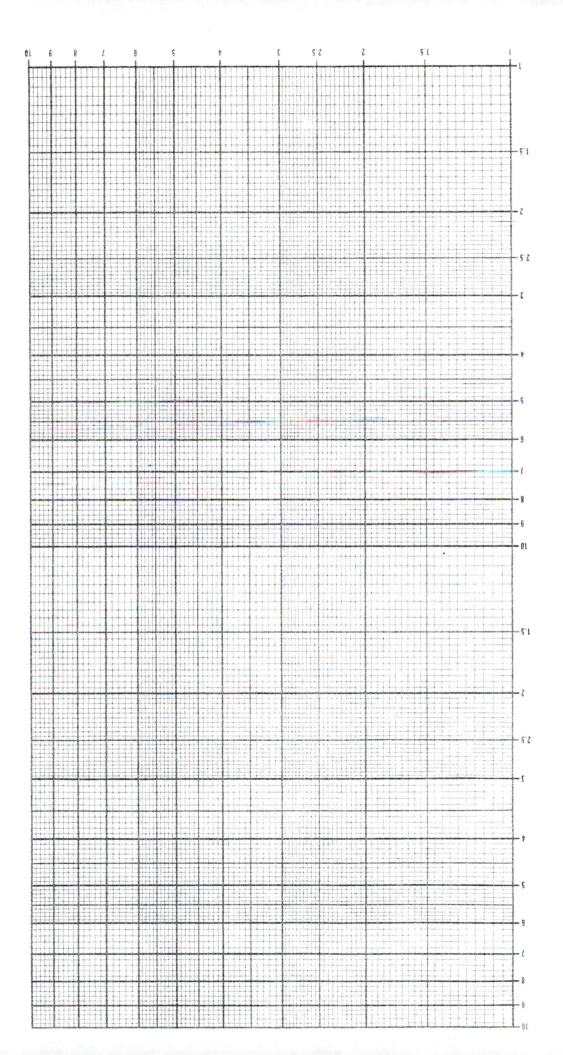

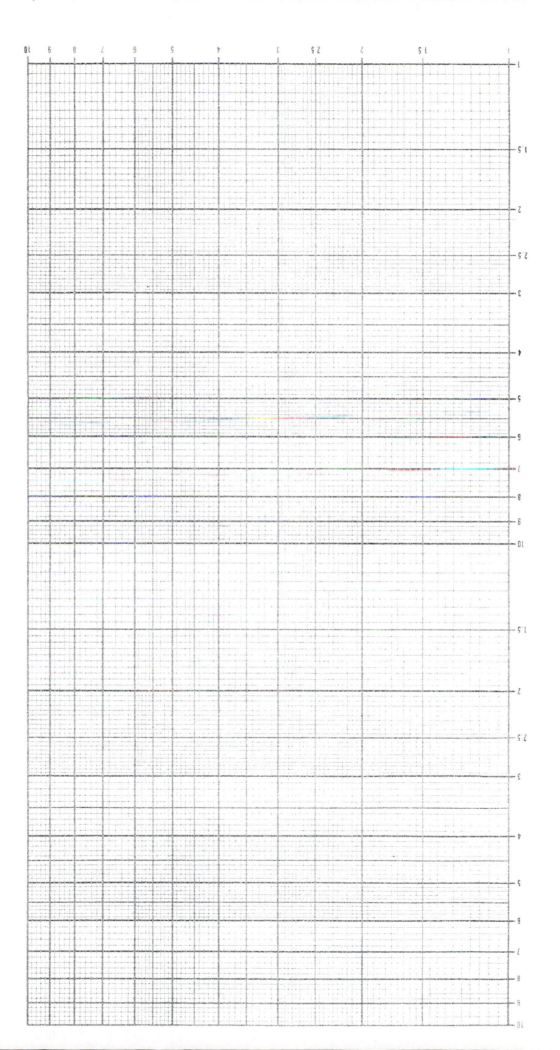

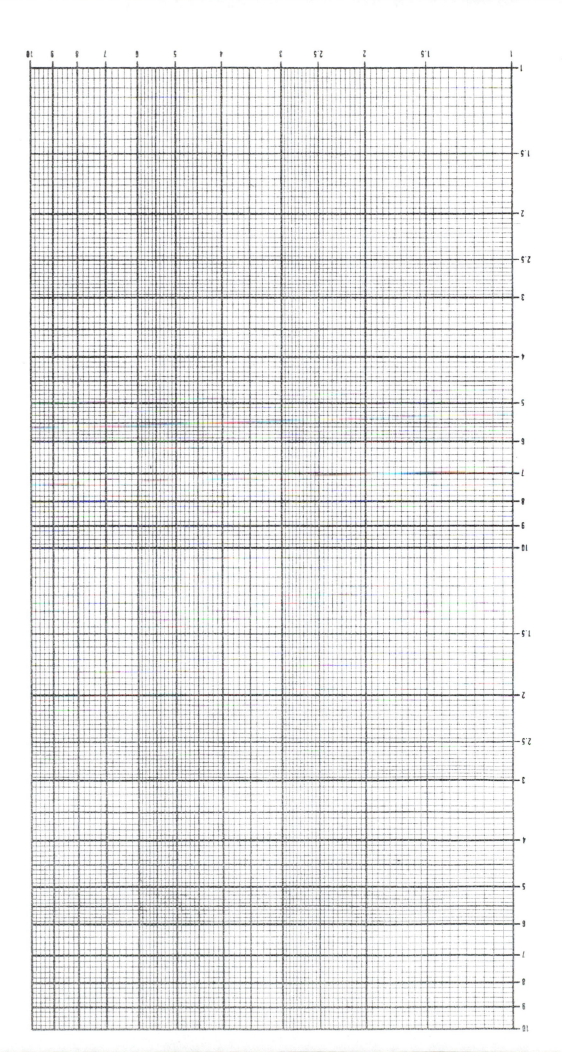

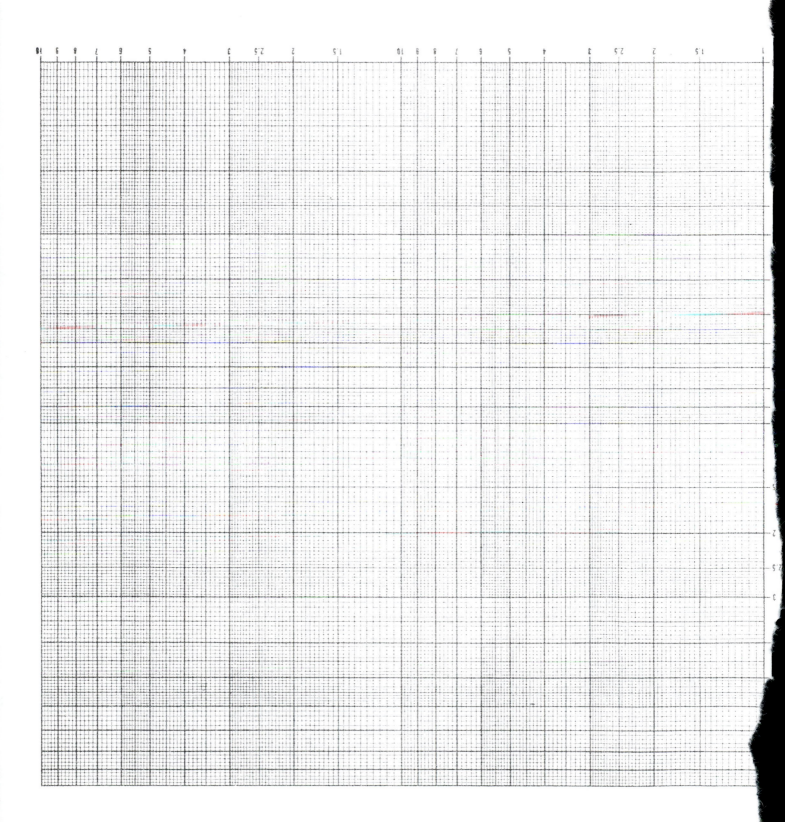

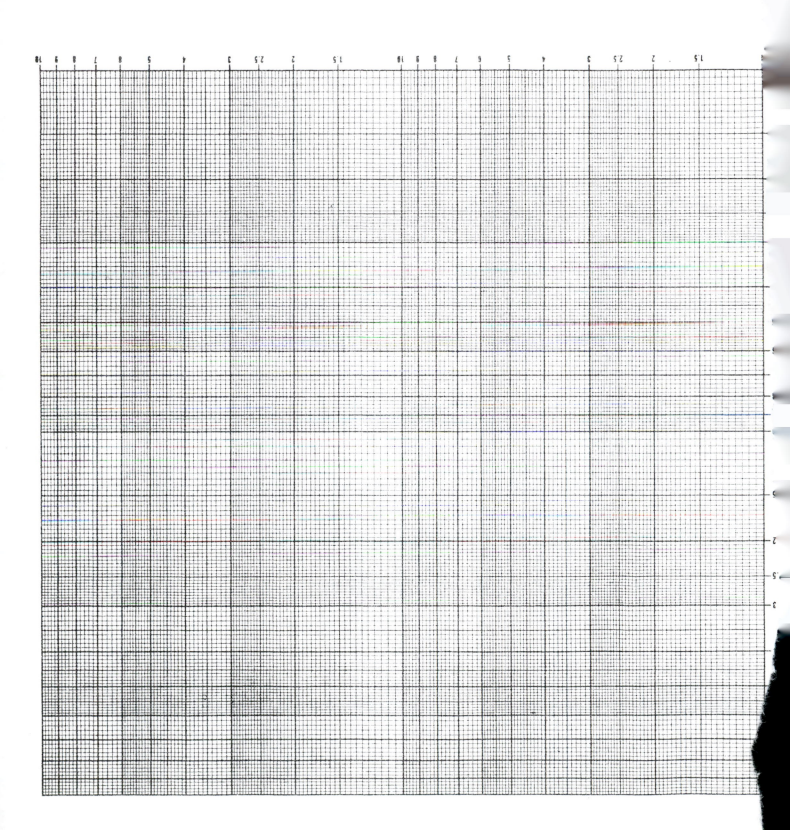

Printed in the United States
6546LVS00004B/83-106